This book series seeks to publish ground-breaking research exploring the productive intersection of science and the cultural imagination. Science is at the centre of daily experience in twenty-first century life and this has defined moments of intense technological change, such as the Space Race of the 1950s and our very own era of synthetic biology. Conceived in dialogue with the field of Science and Technology Studies (STS), this series will carve out a larger place for the contribution of humanities to these fields. The practice of science is shaped by the cultural context in which it occurs and cultural differences are now key to understanding the ways that scientific practice is enmeshed in global issues of equity and social justice. We seek proposals dealing with any aspect of science in popular culture in any genre. We understand popular culture as both a textual and material practice, and thus welcome manuscripts dealing with representations of science in popular culture and those addressing the role of the cultural imagination in material encounters with science. How science is imagined and what meanings are attached to these imaginaries will be the major focus of this series. We encourage proposals from a wide range of historical and cultural perspectives.

More information about this series at
http://www.palgrave.com/gp/series/15760

Alexander Hall

Evolution on British Television and Radio

Transmissions and Transmutations

palgrave
macmillan

Alexander Hall
University of Birmingham
Birmingham, UK

ISSN 2731-4359 ISSN 2731-4367 (electronic)
Palgrave Studies in Science and Popular Culture
ISBN 978-3-030-83042-7 ISBN 978-3-030-83043-4 (eBook)
https://doi.org/10.1007/978-3-030-83043-4

Cover image: Jordan Collver © 2021. This piece was inspired by the Radio Times artwork by Eric Fraser as featured in Chapter 5.

This Palgrave Macmillan imprint is published by the registered company Springer Nature Switzerland AG.
The registered company address is: Gewerbestrasse 11, 6330 Cham, Switzerland

Preface

This book is dedicated to my Grandad, Raymond Darnley, who amongst many other things taught me to always let my curiosity lead me. If you were here to read this, you could see I listened when you said, "Lad, you're going to have to do something repetitive for the rest of your life; it might as well be something that you enjoy."

I began the primary research that informs this book in 2015 as part of the "Science and Religion: Exploring the Spectrum" project funded by the Templeton Religion Trust (TRT0082). In the early stages, I imagined that the project was going to be a symmetrical treatment of evolution in scientific and religious broadcasting in the UK. However, it quickly became apparent that such a lens did not reflect the history of broadcasting on evolution and would not provide much insight. Broadening the scope, what emerged instead was a much more fascinating story about how evolutionary themes have featured across a broad range of genres on both radio and television.

I wrote much of this book during the COVID-19 pandemic of 2020–2021. I would like to highlight that if there is an audience figure missing, or an archival lead that I have not followed up in the way you would like, it is likely that it was due to restrictions in place during the pandemic. At the same time I own any mistakes that have slipped through and make no apologies for the ways that this trying time for humanity have influenced my work, whether for better or for worse.

As ever with these endeavours, although just one name appears on the cover of this book, this project would not have been possible without the input, advice, support and guidance of a large number of people. Firstly, my thanks go to Professors Fern Elsdon-Baker and Bernard Lightman for taking a punt on a fresh-faced postdoc and for continually pushing me and supporting me in equal measure. Thanks to all of those fabulous archivists and librarians who have helped me to piece together the disparate sources that inform this book, in particular the staff at the BBC Written Archives. Huge thanks to the wonderful illustrator, Jordan Collver, for his fabulous cover art. And of course, I must thank all of the friends and colleagues who have given me feedback, whether on early conference papers that became chapters or those who meticulously read final drafts for me. In particular James Riley, Will Mason-Wilkes and everyone else in the Science, Knowledge and Belief in Society Reading Group, Jakob Whitfield, Dominic Berry, James Sumner, Tim Boon, Max Long and of course the indomitable Jessica van Horssen. Jess, thanks so much for all of your support—without both your physical labour and emotional sustenance, this book would have never seen the light of day.

I hope you enjoy reading this book as much as I have enjoyed writing it—piecing together how British broadcasters told the story of evolution across the twentieth century has truly been an honour. Here's to transmissions and transmutations!

Birmingham, UK
June 2021

Alexander Hall

Contents

Abbreviations

AR	Associated-Rediffusion
BAAS	British Association for the Advancement of Science
BBC	British Broadcasting Corporation
BBC/OU	British Broadcasting Corporation and Open University partnership
BBC-WA	British Broadcasting Corporation Written Archives
BSSRS	British Society for Social Responsibility in Science
CRAC	BBC Central Religious Advisory Committee
FEU	BBC Forces Education Unit
ITV	Independent Television
NHU	BBC Natural History Unit
OU	Open University
SF	Science Fiction
SFTV	Science Fiction Television

List of Figures

List of Tables

CHAPTER 1

Introduction

In 2022, the British Broadcasting Corporation (BBC) will premier its latest scientific blockbuster, *Planet Earth III.* Millions of people in the UK and around the world will watch the multimillion-pound natural history documentary made by the famous BBC Natural History Unit. The series is the latest offering in the *Planet Earth* franchise, which began with 2001s groundbreaking *Blue Planet.* Credited with influencing major upsurges in environmental awareness,[1] the franchise situated the natural history of the world, and our underpinning biological understanding of it, within a progressive narrative. In promoting the 2022 series, the BBC stated,

> Combining the awe and wonder of the original Planet Earth, the new science and discoveries of Blue Planet II and Planet Earth II, and the immersive

[1] Matilda Eve Dunn, Morena Mills, and Diogo Veríssimo, 'Evaluating the Impact of the Documentary Series Blue Planet II on Viewers' Plastic Consumption Behaviors', *Conservation Science and Practice* 2, no. 10 (2020): e280, https://doi.org/10.1111/csp2.280; Jennifer Males and Peter Van Aelst, 'Did the Blue Planet Set the Agenda for Plastic Pollution? An Explorative Study on the Influence of a Documentary on the Public, Media and Political Agendas', *Environmental Communication* 15, no. 1 (2 January 2021): 40–54, https://doi.org/10.1080/17524032.2020.1780458.

A. Hall, *Evolution on British Television and Radio*, Palgrave Studies in Science and Popular Culture,
https://doi.org/10.1007/978-3-030-83043-4_1

character-led storytelling of Dynasties, the series will take the 'Planet Earth' experience to new heights.[2]

This narrative framing and the format devices used to support the communication of awe and wonder in the *Planet Earth* series are not limited to BBC natural history broadcasts, appearing across a multitude of BBC science output in this mode. Most notably, in 2010 the resplendent depiction of the world transferred to the cosmic scale, as the physicist Brian Cox introduced viewers to the *Wonders of the Solar System*. Commanding average UK viewing figures of 3.14 million, and widely syndicated and shown all over the world,[3] Cox's awestruck delivery, particularly his over-use of superlatives and the "Five Ws", was quickly parodied.[4] As well as drawing on a tradition of natural theological language, the series and its sequels, *Wonders of the Universe* (2011) and *Wonders of Life* (2013), used what the sociologist Will Mason-Wilkes calls, a "religious" portrayal of science.[5] This includes an explicitly secular humanist vision of humanity, which one BBC Executive Producer described as creating "creation tales for atheists".[6]

[2] Ben Dowell, "The BBC lines up Planet Earth III for 2022", *Radio Times Online*, February 18, 2019, https://www.radiotimes.com/tv/documentaries/planet-earth-3-bbc-2022/ *(last accessed 16/06/2021)*.

[3] Viewing figures are taken from Broadcasters' Audience Research Board Online, https://www.barb.co.uk/viewing-data/ *(last accessed 23/03/21)*. The series aired in among other countries, Australia, Austria, Iceland, the Netherlands, New Zealand, Slovakia and the US. See *Wonders of the Solar System*, Wikipedia, https://en.wikipedia.org/wiki/Wonders_of_the_Solar_System *(last accessed 23/03/21)*.

[4] See, for example, Barry Shitpeas and Philomena Cunk discussing Brian Cox's 'Wonders of Life', segment in *Charlie Brooker's Weekly Wipe*, February 21, 2013, BBC Two.

[5] Will Mason-Wilkes, 'Divine DNA? "Secular" and "Religious" Representations of Science in Nonfiction Science Television Programs', *Zygon*, 55:1 (2020): 6–26, https://doi.org/10.1111/zygo.12574.

[6] Will Mason-Wilkes, 'Science as Religion? Science Communication and Elective Modernism' (PhD, Cardiff University, 2018), http://orca.cf.ac.uk/109735/, 273–277. In addition to his own publicly stated personal humanist beliefs, media commentators quickly began to associate Cox's documentary approach with various secular humanist and atheist movements. See: "BHA welcomes its newest Distinguished Supporter, Professor Brian Cox, as he talks "Big Science" at BHA's annual Voltaire Lecture", *Humanists UK Press Release*, April 8, 2010, https://humanism.org.uk/2010/04/08/news-527/ *(last accessed 23/03/21)*; "The Wonder Years", *The Times*, September 1, 2011, 2; Jonathan Sacks, "Without God we tend to start worshipping ourselves", *The Times*, February 23, 2013, 79; and Eliane Glaser, "Comment: The altar of Cox and co: Many popular scientists are atheist, but they use religious language and cast themselves as priests", *The Guardian*, March 1, 2013, 38.

While the above examples are not directly about evolutionary science, these television series are all set within an evolutionary framework, one that incorporates progressive narratives both of science itself and of humanity more broadly. In illuminating the emergence of this dominant, popular way of speaking about science, this book charts the history of how biological evolution has been depicted, discussed and disseminated via the mediums of television and radio in the UK. Beginning with the very first radio broadcast on evolution, *The Stream of Life* (1925), written and presented by biologist and science populariser, Julian Huxley (1887–1975), the book covers the period up to the 150th anniversary of Charles Darwin's *Origin of the Species* in 2009. This book is the first comprehensive history of how broadcasters have communicated a scientific discipline via radio and television across the twentieth century and beyond. In charting this history, I make the case that the dominant utilisation of a progressive grand evolutionary narrative is an important part of the wider development of standardised and normative ways of speaking about science on broadcast media, which developed during the twentieth century. The book shows how this mode of science broadcasting emerged in the UK because of a specific confluence of expertise in evolutionary science, science communication and public service broadcasting. The later chapters demonstrate how media executives and producers subsequently exported this expertise and approach and sold it around the world. The book explores the potential influence and effect of associating scientific knowledge and practice with other philosophies and worldviews, such as scientific humanism, materialism and more generalised narratives of scientific progressivism.

The Evolutionary Epic and Progressive Narratives in Science

In outlining his approach to making "creation tales for atheists", the BBC Executive Producer in Mason-Wilkes' study described a narrative arc, which begins with the big bang, and leads through the emergence and development of life on earth, to humanity and to our modern technological societies.[7] Explicitly informed by our knowledge of physics and evolutionary biology, this cosmic evolutionary narrative is widely used across

[7] Mason-Wilkes, 273.

popular science media.[8] In his Pulitzer Prize-winning book, *On Human Nature* (1978), the biologist E. O. Wilson called this grand evolutionary narrative the "evolutionary epic".[9] While specific details of the narrative have changed over time, the evolutionary epic was used widely by Darwinian popularisers in the late nineteenth century, and can be traced back even further, to early evolutionary thinkers such as Alexander von Humboldt (1769–1859) and Robert Chambers (1802–1871).[10] Unlike some earlier iterations, Wilson's late twentieth-century version was a distinctly atheistic account, which centred on a scientific materialist worldview. Wilson was building on the ideas of other biologists and popularisers, particularly Julian Huxley who first popularised a form of scientifically informed secular humanism. As outlined in Chap. 2, Huxley's own vision for the future of humanity cannot easily be separated from the media he used to popularise it. Bound up in twentieth-century visions of modernity, which centred on the communicative power of first radio, and later television, it was the historian of science Robert Bud who first drew the connection between Huxley's scientific humanism and early BBC science broadcasts of the 1930s.[11] This book builds on Bud's work, outlining how progressive visions of humanity and the evolutionary epic influenced BBC productions, extending the chronology to explore how this narrative became embedded in the formats themselves.

The evolutionary epic is just one synthetic progressive narrative centred on science that came into vogue during the last decades of the twentieth century. Often idiosyncratic and altering in their specifics, historians have only just begun to untangle the myriad ways that twentieth-century progressive intellectuals used scientific narratives in their visions for the future

[8] Jon Turney, 'Telling The Facts Of Life: Cosmology and the Epic of Evolution', *Science as Culture* 10, no. 2 (1 June 2001): 225–47, https://doi.org/10.1080/09505430120052301.

[9] Edward O. Wilson, *On Human Nature* (Harvard University Press, 1978): 201.

[10] James A. Secord, *Victorian Sensation: The Extraordinary Publication, Reception, and Secret Authorship of Vestiges of the Natural History of Creation* (University of Chicago Press, 2003); Bernard Lightman, *Victorian Popularizers of Science: Designing Nature for New Audiences* (University of Chicago Press, 2009); Ian Hesketh, 'The Recurrence of the Evolutionary Epic', *Journal of the Philosophy of History* 9, no. 2 (1 January 2015): 196–219, https://doi.org/10.1163/18722636-12341300; Andrea Wulf, *The Invention of Nature: Alexander Von Humboldt's New World* (Alfred A. Knopf, 2015), 235–248.

[11] Robert Bud, '"The Spark Gap Is Mightier than the Pen": The Promotion of an Ideology of Science in the Early 1930s', *Journal of Political Ideologies* 22:2 (2017): 169–81, https://doi.org/10.1080/13569317.2017.1298548.

of humanity. In *A Final Story: Science, Myth and Beginnings* (2017) Nasser Zakariya provides the most in-depth exploration of scientific epics, skilfully helping us to see the links between scientific knowledge, the human need for storytelling and historical methods. Zakariya connects the major science popularisers of the twentieth century, several of whom feature in this book, with previous generations of universalists and synthetic thinkers back to the Enlightenment. In *Progress Unchained* (2021), Peter Bowler reins in this history to focus on progress in relation to evolutionary biology. In charting a shift from linear to open-ended visions of progress, Bowler highlights how developments in evolutionary science have always been integrated with, and influenced by, wider progressive narratives and societal influences.[12]

Synthetic Narratives in the History of Biology

Any readers familiar with the history of biology in the twentieth century will be aware that progressive visions of science have not merely been tools used to promote and popularise the discipline. As the historian Vassiliki Betty Smocovitis (1996) has demonstrated, biologists involved in the evolutionary synthesis of the 1930s and 1940s purposely sought to use evolution to unify the biological sciences. Smocovitis argues that rather than simply being a synthesis between Darwin's natural selection and genetics, proponents of the evolutionary synthesis were part of wider philosophical trends that sought to unify knowledge.[13] While this book does not aim to analyse in-depth the scientific content of broadcasts on evolution, without the earlier work of Smocovitis and other historians of biology, such as

[12] For a striking case study of how geopolitical fears infused scientific discussions and scientists utilised popular mediums and narratives to bypass the mechanisms of institutional science, and in doing so popularised controversial ideas about violence among apes, which were often extrapolated to theorise on the emergence of aggression in humans, see Erika Lorraine Milam, *Creatures of Cain: The Hunt for Human Nature in Cold War America* (Princeton University Press, 2019).

[13] Vassiliki Betty Smocovitis, *Unifying Biology: The Evolutionary Synthesis and Evolutionary Biology* (Princeton University Press, 1996). For a UK-focused history of the evolutionary synthesis, see: Maurizio Esposito, 'Utopianism in the British Evolutionary Synthesis', *Studies in History and Philosophy of Science Part C: Studies in History and Philosophy of Biological and Biomedical Sciences*, Defining Darwinism: One Hundred and Fifty Years of Debate, 42:1 (2011): 40–49, https://doi.org/10.1016/j.shpsc.2010.11.007.

Peter Bowler (1989), my focus on the narrative form of broadcasts on the subject would not be possible.[14]

Media Histories and Studies of Science

While media histories of science have grown in quantity and scope in recent years, they are still small in number.[15] In particular, there are still few histories of science on the radio.[16] Where there are gaps, especially in regards to the UK, this book draws on more general histories of British broadcasting, principally the institutional histories of Asa Briggs and Jean Seaton, and foundational media histories by scholars such as Paddy Scannell and Andrew Crisell.[17] In writing on the history of British science media, my work draws heavily on the scholarship of historians of popular science Timothy Boon, Allan Jones, Peter Bowler and Jean-Baptiste Gouyon.[18] Boon's *Films of Fact* (2008) remains the most comprehensive treatment of science in British documentaries, while Gouyon's recent

[14] Peter J. Bowler, *Evolution: The History of an Idea* (University of California Press, 1989).

[15] A notable non-UK focused history of science media that has informed this book is Marcel Chotkowski LaFollette, *Science on American Television: A History* (University of Chicago Press, 2013).

[16] Both Bud, '"The Spark Gap Is Mightier than the Pen"'; and Allan Jones, 'Science in the Making: 1930s Citizen Science on the BBC', *History of Education*, 49:3 (2020): 1–17, deal with a limited selection of science radio broadcasts from the 1930s. While a history of the earlier science popularisers on American radio is provided by Marcel Chotkowski LaFollette, *Science on the Air: Popularizers and Personalities on Radio and Early Television* (Chicago: University of Chicago Press, 2009).

[17] See Asa Briggs, *The History of Broadcasting in the United Kingdom: Volume V: Competition* (Oxford: Oxford University Press, 1995); Jean Seaton, *Pinkoes and Traitors: The BBC and the Nation, 1974–1987* (Profile Books, 2015); Paddy Scannell and David Cardiff, *A Social History of British Broadcasting: Volume 1 – 1922–1939, Serving the Nation* (Wiley, 1991); and Andrew Crisell, *An Introductory History of British Broadcasting* (Psychology Press, 2002).

[18] For illustrative examples of their work see: Tim Boon, '1962: "What Manner of Men?": Meeting Scientists through Television', *Public Understanding of Science*, 2018, https://doi.org/10.1177/0963662518805314; Allan Jones, 'Science, the BBC and the Two Cultures', in *Broadcasting in the UK and US in the 1950s: Historical Perspectives*, ed. Jamie Medhurst, Sian Nicholas, and Tom O'Malley (Cambridge Scholars Publishing, 2016), 101–20; Peter J. Bowler, *Science for All: The Popularization of Science in Early Twentieth-Century Britain* (University of Chicago Press, 2009); Jean-Baptiste Gouyon, 'From Engaged Citizen to Lone Hero: Nobel Prize Laureates on British Television, 1962–2004', *Public Understanding of Science* 27:4 (2018): 446–57, https://doi.org/10.1177/0963662518760790.

study of BBC natural history documentaries has greatly informed this book (Chap. 3). This book extends beyond these texts, and their largely institutional focus, exploring the role that progressive evolutionary narratives have played in popular science broadcasts. By focusing in on one scientific subject, I am able to analyse both radio and television broadcasts over a longer period and across multiple genres. In extending my focus beyond non-fiction science broadcasting, scholarship exploring science's depiction in movies and science fiction television has proven extremely insightful: most notably, the work of David Kirby on the history of evolutionary narratives in Hollywood and on the relationship between scientific accuracy and plausibility in science fiction (Chap. 5).[19]

Humanist Blockbusters

In this book, I trace the history of the secular humanist portrayal of science and its utilisation of the evolutionary epic, as found in contemporary BBC science series such as the *Wonders* franchise, back to its earliest proponents. Elsewhere, I have argued that this approach, exemplified by Jacob Bronowski's thirteen-part series, *The Ascent of Man* (1973), is distinct enough to warrant its own categorisation. I use the term "humanist blockbusters" to refer to these broadcasts, which are typically big-budget documentaries that use spectacular visuals and the language of awe to present a grand narrative centred on the progressive nature of scientific discovery.[20] Aligned somewhat with Derek Bousé's "blue-chip" wildlife films, and reflecting broader trends in science documentaries, humanist

[19] David A. Kirby, *Lab Coats in Hollywood: Science, Scientists, and Cinema* (MIT Press, 2011); David A. Kirby, 'Harnessing the Persuasive Power of Narrative: Science, Storytelling, and Movie Censorship, 1930–1968', *Science in Context*, 31:1 (2018): 85–106, https://doi.org/10.1017/S0269889718000029; and David A. Kirby, 'The Devil in Our DNA: A Brief History of Eugenics in Science Fiction Films', *Literature and Medicine* 26:1 (2007): 83–108, https://doi.org/10.1353/lm.2008.0006. For introductory texts on science fiction and science fiction television see: Mark Bould and Sherryl Vint, *The Routledge Concise History of Science Fiction* (Routledge, 2011); and J. P. Telotte, *The Essential Science Fiction Television Reader* (University Press of Kentucky, 2008).

[20] Alexander Hall, 'A Humanist Blockbuster? Jacob Bronowski and The Ascent of Man', in *Rethinking History, Science, and Religion: An Exploration of Conflict and the Complexity Principle*, ed. Bernard Lightman (University of Pittsburgh Press, 2019).

blockbusters are usually filmed in multiple locations, funded by major co-production agreements and rely on cutting-edge technologies.[21]

In Chaps. 2 and 3, I show how early evolutionary broadcasts by Julian Huxley influenced Jacob Bronowski's earliest radio broadcasts in the 1940s. Both used a distinct approach to science broadcasting that focused on science in society, an approach developed by pioneering producers at the BBC such as Mary Adams.[22] In Chap. 6, we see how Bronowski distilled the thesis of his humanist blockbuster, *The Ascent of Man* from his earlier broadcasts and popular writings, but also that its success relied on the expertise of BBC features producers, most notably Adrian Malone. Extending the narrative beyond 1973, the chapter explores how the production team who worked on *Ascent* used the devices, format and narrative developed for the series on other major projects, such as David Attenborough's *Life on Earth* (1979) and Carl Sagan's *Cosmos: A Personal Voyage* (1980). The preceding chapters (Chaps. 4 and 5) chart how the narrative form and televisual devices used in *Ascent* were developed by producers across a range of departments at the BBC. In doing so, I go beyond earlier scholarship, which focuses on synthetic popularisers and their progressive philosophies (e.g. Bowler, 1989; Lightman, 2009; Zakariya, 2017) to interrogate the media formats used to communicate evolutionary ideas. This expanded focus, which brings into view the intellectuals behind the scenes who created, refined and shaped the narrative and its audio-visual aids, is essential if we are to understand how evolutionary ideas and their communication influenced sciences shifting role in society.[23]

This book demonstrates that the general progressive narrative used in *Ascent* was only one of several approaches that British broadcasters have used to communicate evolutionary science and evolutionary themes. Further, we see that its emergence as the dominant and default framing for both content on evolution and broadcasts on science in society more

[21] Derek Bousé, *Wildlife Films* (University of Pennsylvania Press, 2011), 14–22; Jean Seaton, *Pinkoes and Traitors: The BBC and the Nation, 1974–1987* (Profile Books, 2015); Timothy Boon, *Films of Fact: A History of Science in Documentary Films and Television* (Wallflower Press, 2008).

[22] For more on Adams' career see: Allan Jones, 'Mary Adams and the Producer's Role in Early BBC Science Broadcasts', *Public Understanding of Science* 21:8 (2012): 968–83, https://doi.org/10.1177/0963662511419450.

[23] Bowler, *Evolution*; Lightman, *Victorian Popularizers of Science*; and Nasser Zakariya, *A Final Story: Science, Myth, and Beginnings* (University of Chicago Press, 2017).

generally was not inevitable. In Chaps. 2 and 3 I show how early science broadcasts on evolution often featured accommodationist religious figures who in later decades were completely absent from science broadcasts on the subject. In Chaps. 4 and 5, we see how it was the more experimental spaces of educational broadcasts and radio dramas that first developed the historical re-enactment through which Charles Darwin became a touchstone and leitmotif of a wide range of broadcasts with evolutionary themes. In addition to more in-depth historical broadcasts (Chaps. 7 and 8), the Darwin flashback was used by science broadcasters to contextualise contemporary evolutionary developments, to frame conflicts over evolution as historic phenomena, and to create an exalted position for Darwin in the mythic account of evolution (Chap. 8). This mythic account was part of a wider "Darwin Industry", which by the widely celebrated 2009 Darwin anniversary had largely erased the contribution of Darwin's contemporaries, particularly the co-discoverer of evolution by natural selection, Alfred Russel Wallace (1823–1913) (Chap. 8).

Alongside celebratory and progressive modes, science broadcasting in the UK has occasionally provided space for critiques of science. Yet science broadcasts that have featured critical discussion of science have almost always centred on socio-political aspects or specific contested subject areas, rather than providing space for the criticism of scientific process or scientific hegemony itself (Chap. 6). In keeping with this, from the 1920s until the 1980s, British broadcasters provided little airtime to anti-evolution arguments. However, during the early 1980s broadcasters began covering the resurgence of creationist movements in the USA. Beginning in a journalistic mode that simply attempted to report trends in the US, science broadcasts began including creationism and the threat of the movement's growth in the UK as a hook to attract an audience. These programmes presented evolutionary science as an explanatory system that was superior to fundamentalist religious traditions, and reinforced the boundary between proper science and anti-evolutionist pseudoscience (Chap. 7).

Approaches to Studying Media

Informed by a wide range of media studies, the approach adopted in this book attempts to explore the "full circuit of mass communication".[24] Situating broadcasts within their cultural and institutional contexts, I explore them from their initial conception, through their production, distribution, audience reception and lasting influence. Roger Silverstone's seminal ethnography of the making of a BBC science documentary, *Framing Science*, remains the most in-depth study of the production of BBC science content. Although as a historical study this book includes more contextual content, Silverstone's meticulous exploration of how the specific processes of making a science documentary shaped the framing of the subject matter has influenced my approach significantly.[25] More recently, Vincent Campbell's *Science, Entertainment and Television Documentary* (2016) explored contemporary British science broadcasts as factual entertainment, focusing on identifying narrative structures, visual tropes and other devices used to frame the subject matter. Adding a longitudinal lens to Campbell's approach by exploring the contingent emergence of narrative structures and visual devices used to communicate evolutionary themes over many decades, I expose how a dominant or normative frame on evolutionary science was established. Throughout, I follow Entman's (1993) definition, that framing involves selection and salience, and attempts to expose editorial decisions that "promote a particular problem definition, causal interpretation, moral evaluation, and/or treatment recommendation" of the subject matter.[26] However like Campbell (2016), this study follows a more qualitative tradition, and does not attempt to use the concept of media frames as part of a systematic content analysis as used in other media studies.[27] In this study, I analyse media frames across different genres to make visible past decisions on presenting scientific knowledge and ideas in a particular way. Normative

[24] David Miller, *The Circuit of Mass Communication: Media Strategies, Representation and Audience Reception in the AIDS Crisis* (SAGE, 1998), 9.

[25] Roger Silverstone, *Framing Science: The Making of a BBC Documentary* (BFI Publishing, 1985).

[26] Robert M. Entman, 'Framing: Toward Clarification of a Fractured Paradigm', *Journal of Communication* 43:4 (1993): 52, https://doi.org/10.1111/j.1460-2466.1993.tb01304.x.

[27] For an overview of the major approaches to content analysis with regard to media frames see: Jörg Matthes and Matthias Kohring, 'The Content Analysis of Media Frames: Toward Improving Reliability and Validity', *Journal of Communication* 58:2 (2008): 258–79, https://doi.org/10.1111/j.1460-2466.2008.00384.x.

frames which today are so embedded in the formats themselves they are often difficult to discern.

Influenced by both historical and media studies on audience reception and influence,[28] and sociological studies on perceptions of evolution in British society,[29] in researching this book it quickly became apparent that while popular media on the subject shaped individual perceptions, no one, individual series alone significantly informed viewers. We can see the influence that media on evolutionary themes has had on British society, not only with regard to the dissemination of scientific knowledge, but also in relation to the popularity of progressive narratives.[30] Big-budget documentaries, such as David Attenborough's *Life on Earth* (1979), were only one element of a tapestry of broadcast media that individual audience members consumed. Further, broadcast media is just one influence among many, from school and university education through to familial and religious traditions that influences knowledge on, and acceptance of evolution.[31] Therefore, rather than just focusing on the dissemination of biological evolutionary knowledge in a deficit model tradition of science communication,[32] this book considers evolutionary themes across all genres of radio and television broadcasting in the UK. By expanding from

[28] For example, Secord, *Victorian Sensation*; and Miller, *The Circuit of Mass Communication*.

[29] Fern Elsdon-Baker, 'Creating Creationists: The Influence of "Issues Framing" on Our Understanding of Public Perceptions of Clash Narratives between Evolutionary Science and Belief', *Public Understanding of Science* 24:4 (2015): 422–39, https://doi.org/10.1177/0963662514563015; Stephen H. Jones, Rebecca Catto, and Tom Kaden, eds., *Science, Belief and Society: International Perspectives on Religion, Non-Religion and the Public Understanding of Science*. (University of Bristol Press, 2019).

[30] Fern Elsdon-Baker and Bernard Lightman, *Identity in a Secular Age: Science, Religion, and Public Perception* (University of Pittsburgh Press, 2020); Amy Unsworth and David Voas, 'The Dawkins Effect? Celebrity Scientists, (Non)Religious Publics and Changed Attitudes to Evolution', *Public Understanding of Science* (2021), https://doi.org/10.1177/0963662521989513.

[31] Thomas Aechtner, 'Galileo Still Goes to Jail: Conflict Model Persistence Within Introductory Anthropology Materials', *Zygon* 50:1 (2015), 209–26, https://doi.org/10.1111/zygo.12149; Stephen H. Jones et al., '"That's How Muslims Are Required to View the World": Race, Culture and Belief in Non-Muslims' Descriptions of Islam and Science', *The Sociological Review* 67:1 (2019), 161–77, https://doi.org/10.1177/0038026118778174; and Thomas Aechtner, 'Darwin-Skeptic Mass Media: Examining Persuasion in the Evolution Wars', *Journal of Media and Religion* 13:4 (2014), 187–207, https://doi.org/10.1080/15348423.2014.971559.

[32] The deficit model describes an approach to science communication that assumes lack of interest, support or belief in a scientific subject is caused simply by ignorance or lack of knowledge. Jane Gregory and Steven Miller, *Science In Public* (Basic Books, 2000), Chap. 4.

the term evolutionary science(s) to broader evolutionary themes, my aim is to capture the multiplicity of ways that evolutionary ideas have been used to forward philosophical, political and social agendas that extend well beyond the bounds of the biological sciences. As the following chapters attest, others areas of society have repeatedly co-opted biological evolutionary concepts, from natural selection to genetic determinism.

When we analyse audience reception and influence alongside the production history of broadcasts and the philosophical frameworks of science popularisers, some interesting patterns emerge. For example, the most popular primetime content with evolutionary themes, whether blockbuster documentaries (Chap. 6) or science fiction thrillers (Chap. 5), were the most reductive. Dramatisations and fantastical science fiction often featured familiar tropes, such as Darwinian conflict, or immoral evolutionary scientists. However, these fictional worlds also allowed more space than factual content for discussion of uncertainty, gaps in knowledge, future directions and other philosophical questions related to evolution. Educational broadcasts, on the other hand, rarely gave space to discuss such themes. Instead, they have taken a more didactic approach and focused on teaching evolution within a progressive framework (Chap. 4). Whether topical science serials or major documentary series, many factual broadcasts on evolution have also relied on reductive tropes, using the evolutionary epic to reinforce the hegemonic authority of the biological sciences, and define the boundaries of science (Chap. 7). Appealing to an "in-group" audience who already consider science an important component of their cultural identity, these factual programmes have reinforced and promoted evolutionary sciences' position at the centre of the march of scientific progress.

Surveying British Broadcasting on Evolution

When embarking on this project I set out to write a book that was more than an institutional history of evolution at the BBC. This is reflected in my attempts to demonstrate that many science producers and broadcast educationalists were part of the same social milieu of more well-known science popularisers (Chap. 3) and literary intellectuals (Chap. 5). Nonetheless, as you will see in the following pages, the BBC dominates proceedings. This is for several reasons, some of which relate to

For more on the deficit model in relation to broadcasts on creationism and evolution see Chap. 7.

the pragmatic challenges of archival research. Radio and television are by design ephemeral media; archiving of transcripts, production files and other records has been inconsistent, particularly in the shifting landscape of British commercial media franchises. However, the main reason is due to the dominance of the BBC, which held a monopoly on British television broadcasting until 1955 and on radio until 1973.[33] Despite a brief window of educational broadcasting on ITV in the late 1950s and early 1960s (Chap. 4), and a small stream of documentaries on Channel 4 following its launch in 1982 (Chaps. 7 and 8), the overwhelming majority of broadcasts with evolutionary themes have been on one of the BBC's many channels.

Setting out with the aim of covering all nationally broadcast content on biological evolution in British broadcasting history, my research for this project began with an in-depth survey of the long-running listings magazine, the *Radio Times*. Using the digitised platform of BBC listings from the magazine for the period from 1923 to 2009, I searched all content containing the word evolution and manually filtered results, removing all those that used the word in a non-biological context.[34] This survey returned a corpus of 1167 individual episodes, 380 of which were on radio and 787 on television, which increased in frequency across the century, approximately in line with increases in the BBC's total broadcasting output (Fig. 1.1).[35] Analysis of this top-level data provided an indicative

[33] For more on the launch of Independent Television (ITV) in 1955 see Asa Briggs, *The History of Broadcasting in the United Kingdom: Volume IV: Sound and Vision* (Oxford University Press, 1995); and Bernard Sendall, *Independent Television in Britain: Origin and Foundation 1946–62* (Springer, 1982). Despite the ongoing popularity of offshore pirate radio stations, it was not until the autumn of 1973, following the passing of the Sound Broadcasting Act (1972) that the first legal independent radio stations were launched. Briggs, *The History of Broadcasting in the United Kingdom*, 1995, 885–89.

[34] The original listings for all of the BBC show referenced throughout this book can be found online at the digitised version of the listings magazine the *Radio Times*. BBC Genome Project, http://genome.ch.bbc.co.uk/ *(last accessed 16/06/21)*. The descriptions featured in the listings were written by the individual production teams creating each show, and therefore provide a relatively accurate proxy for surveying BBC content across this period. For more on the history of the magazine see Tony Currie, *The Radio Times Story* (Kelly Publications, 2001). All BBC copyright content reproduced courtesy of the British Broadcasting Corporation. All rights reserved.

[35] A version of this corpus focused on the years 1945–2009 first appeared in Alexander Hall, 'Evolution on the Small Screen: Reflections on Media, Science, and Religion in Twentieth-Century Britain', in *Identity in a Secular Age: Science, Religion, and Public Perceptions*, ed. Fern Elsdon-Baker and Bernard Lightman (University of Pittsburgh Press, 2020).

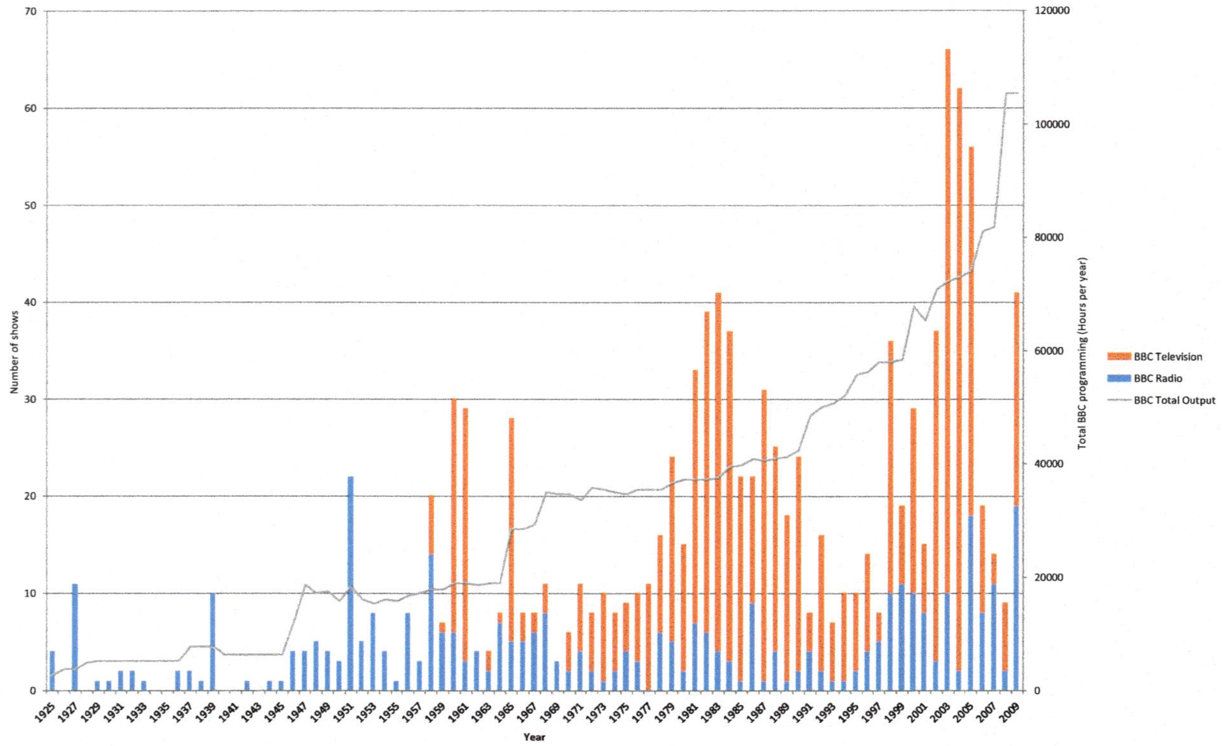

Fig. 1.1 Number of radio and television broadcasts mentioning "evolution" in a biological context as listed in the BBC *Radio Times* by year (1925–2009). Secondary axis approximates total output of BBC radio and television combined in hours per year

insight, allowing me to explore changes in genres over time, and flagging other notable trends such as spikes in coverage around significant anniversaries (Chap. 8).

Building from these results, I used snowball-sampling methods and comprehensively explored all of the other broadcasts produced by the departments, producers and writers returned in the corpus. I then used newspaper listings and other grey literature to expand beyond just BBC-produced content. The book draws primarily on archival research conducted at the BBC Written Archives in Reading (BBC-WA), supplemented by records from the ITV Archive in Leeds, and other personal and institutional archives both physical and digital. In addition to these primary materials and secondary literature from a wide range of disciplines, I have also drawn on a vast wealth of grey literature, including BBC pamphlets, published versions of radio and television series, and for the latter period, recordings of the programmes themselves.

Summary of the Argument

While structured largely chronologically, the book is more than a straightforward history of science broadcasting. It explores a diverse range of genres that have broadcast content on evolution, from science documentaries to science fiction. In thematically dealing with each of these genres of radio and television, I analyse how broadcasters communicated the science of evolution to generations of audiences in the UK. While the book covers the specific aspects of evolutionary theory that have been communicated via the airwaves, such as when the modern evolutionary synthesis was first disseminated, a reductive focus on the scientific content of past media productions is not my primary focus. Rather, I show how producers conceived, constructed and communicated content on evolutionary themes to various audiences across the UK. In doing so, these chapters explore how media technologies and popular narratives of progress shaped the communication of one specific scientific subject, and in turn, how over time normative framings, narrative structures and visual devices embedded the approaches in the various formats themselves. By utilising a wide range of sources and situating each show in its contingent cultural context, the book aims to show how broadcasters and popularisers have told the story of evolution, and in turn how these depictions have influenced and interacted with wider understandings of evolution, at both the individual and societal level.

Chapter 2 begins with *The Stream of Life* (1925), a series of short radio lectures by Julian Huxley. This was the first time radio broadcasters introduced British listeners to evolution. While pressing concerns of the day, such as "heredity and racial problems", influenced the content, the lectures also introduced audiences to Huxley's own ideology on the importance of evolution for the future of humanity.[36] From here on in, such scientific and evolutionary-humanist ideas, as first espoused on air by Huxley, repeatedly re-emerged across multiple decades, departments and genres of British broadcasting. First Huxley, and then subsequently his contemporary Bronowski, emerged as the main popularisers of this worldview. If this book were a more conventional history of ideas, Huxley and Bronowski would be our protagonists. However, given that my central argument extends from a commitment to exploring the full circuit of mass communication and its influence on society, the cast introduced over the following chapters is much more diverse. Without which, we cannot truly understand how progressive visions of science, particularly those within an evolutionary framework, have transferred from one ideological vision of the future to the dominant way that popular media speak about science and society today. Thus over the coming chapters, an array of historical figures from behind the camera and microphone, and the pressures shaping the broadcasts they produced, are given an equal footing with their more famous academic contemporaries. From those with a frustratingly small historical footprint, such as Nesta Pain (1905–1995) who first prioritised dramatic values in science broadcasts, through to those who rose to become senior executives, such as Aubrey Singer (1927–2007), this shift in focus aims to broaden our understanding of how mass media amplifies, refines, embeds and normalises philosophical worldviews in the popular conscience.

References

Aechtner, Thomas. 'Darwin-Skeptic Mass Media: Examining Persuasion in the Evolution Wars'. *Journal of Media and Religion* 13:4 (2014): 187–207. https://doi.org/10.1080/15348423.2014.971559.

———. 'Galileo Still Goes to Jail: Conflict Model Persistence Within Introductory Anthropology Materials'. *Zygon* 50:1 (2015): 209–26. https://doi.org/10.1111/zygo.12149.

[36] Julian Huxley, *The Stream of Life* (Watts & Co., 1926).

"BHA welcomes its newest Distinguished Supporter, Professor Brian Cox, as he talks "Big Science" at BHA's annual Voltaire Lecture", *Humanists UK Press Release*, April 8, 2010, https://humanism.org.uk/2010/04/08/news-527/

Boon, Tim. '1962: "What Manner of Men?": Meeting Scientists through Television'. *Public Understanding of Science* (2018). https://doi.org/10.1177/0963662518805314.

Boon, Timothy. *Films of Fact: A History of Science in Documentary Films and Television*. Wallflower Press, 2008.

Bould, Mark, and Sherryl Vint. *The Routledge Concise History of Science Fiction*. Routledge, 2011.

Bousé, Derek. *Wildlife Films*. University of Pennsylvania Press, 2011.

Bowler, Peter J. *Evolution: The History of an Idea*. University of California Press, 1989.

———. *Science for All: The Popularization of Science in Early Twentieth-Century Britain*. University of Chicago Press, 2009.

Briggs, Asa. *The History of Broadcasting in the United Kingdom: Volume IV: Sound and Vision*. OUP Oxford, 1995.

———. *The History of Broadcasting in the United Kingdom: Volume V: Competition*. OUP Oxford, 1995b.

Bud, Robert. '"The Spark Gap Is Mightier than the Pen": The Promotion of an Ideology of Science in the Early 1930s'. *Journal of Political Ideologies* 22:2 (2017): 169–81. https://doi.org/10.1080/13569317.2017.1298548.

Crisell, Andrew. *An Introductory History of British Broadcasting*. Psychology Press, 2002.

Currie, Tony. *The Radio Times Story*. Kelly Publications, 2001.

Dowell, Ben. "The BBC lines up Planet Earth III for 2022", *Radio Times Online*, February 18, 2019, https://www.radiotimes.com/tv/documentaries/planet-earth-3-bbc-2022/

Dunn, Matilda Eve, Morena Mills, and Diogo Veríssimo. 'Evaluating the Impact of the Documentary Series Blue Planet II on Viewers' Plastic Consumption Behaviors'. *Conservation Science and Practice* 2:10 (2020). https://doi.org/10.1111/csp2.280.

Elsdon-Baker, Fern. 'Creating Creationists: The Influence of "Issues Framing" on Our Understanding of Public Perceptions of Clash Narratives between Evolutionary Science and Belief'. *Public Understanding of Science* 24:4 (2015): 422–39. https://doi.org/10.1177/0963662514563015.

Elsdon-Baker, Fern, and Bernard Lightman. *Identity in a Secular Age: Science, Religion, and Public Perception*. University of Pittsburgh Press, 2020.

Entman, Robert M. 'Framing: Toward Clarification of a Fractured Paradigm'. *Journal of Communication* 43:4 (1993): 51–58. https://doi.org/10.1111/j.1460-2466.1993.tb01304.x.

Esposito, Maurizio. 'Utopianism in the British Evolutionary Synthesis'. *Studies in History and Philosophy of Science Part C: Studies in History and Philosophy of Biological and Biomedical Sciences*, Defining Darwinism: One Hundred and Fifty Years of Debate, 42:1 (2011): 40–49. https://doi.org/10.1016/j.shpsc.2010.11.007.

Glaser, Eliane. "Comment: The altar of Cox and co: Many popular scientists are atheist, but they use religious language and cast themselves as priests", *The Guardian*, March 1, 2013, 38.

Gouyon, Jean-Baptiste. 'From Engaged Citizen to Lone Hero: Nobel Prize Laureates on British Television, 1962–2004'. *Public Understanding of Science* 27:4 (2018): 446–57. https://doi.org/10.1177/0963662518760790.

Gregory, Jane, and Steven Miller. *Science In Public.* Basic Books, 2000.

Hall, Alexander. 'A Humanist Blockbuster? Jacob Bronowski and The Ascent of Man'. In *Rethinking History, Science, and Religion: An Exploration of Conflict and the Complexity Principle*, edited by Bernard Lightman. University of Pittsburgh Press, 2019.

———. 'Evolution on the Small Screen: Reflections on Media, Science, and Religion in Twentieth-Century Britain'. In *Identity in a Secular Age: Science, Religion, and Public Perceptions*, edited by Fern Elsdon-Baker and Bernard Lightman. University of Pittsburgh Press, 2020.

Hesketh, Ian. 'The Recurrence of the Evolutionary Epic'. *Journal of the Philosophy of History* 9:2 (2015): 196–219. https://doi.org/10.1163/18722636-12341300.

Huxley, Julian. *The Stream of Life.* Watts & Co., 1926.

Jones, Allan. 'Mary Adams and the Producer's Role in Early BBC Science Broadcasts'. *Public Understanding of Science* 21:8 (2012): 968–83. https://doi.org/10.1177/0963662511419450.

———. 'Science in the Making: 1930s Citizen Science on the BBC'. *History of Education* 49:3 (2020): 1–17. https://doi.org/10.1080/0046760X.2020.1712620.

———. 'Science, the BBC and the Two Cultures'. In *Broadcasting in the UK and US in the 1950s: Historical Perspectives*, edited by Jamie Medhurst, Sian Nicholas, and Tom O'Malley, 101–20. Cambridge Scholars Publishing, 2016. http://oro.open.ac.uk/42572/.

Jones, Stephen H., Rebecca Catto, and Tom Kaden, eds. *Science, Belief and Society: International Perspectives on Religion, Non-Religion and the Public Understanding of Science.* University of Bristol Press, 2019a.

Jones, Stephen H., Rebecca Catto, Tom Kaden, and Fern Elsdon-Baker. '"That's How Muslims Are Required to View the World": Race, Culture and Belief in Non-Muslims' Descriptions of Islam and Science'. *The Sociological Review* 67:1 (2019b): 161–77. https://doi.org/10.1177/0038026118778174.

Kirby, David A. 'Harnessing the Persuasive Power of Narrative: Science, Storytelling, and Movie Censorship, 1930–1968'. *Science in Context* 31:1 (2018): 85–106. https://doi.org/10.1017/S0269889718000029.

———. *Lab Coats in Hollywood: Science, Scientists, and Cinema*. MIT Press, 2011.

———. 'The Devil in Our DNA: A Brief History of Eugenics in Science Fiction Films'. *Literature and Medicine* 26:1 (2007): 83–108. https://doi.org/10.1353/lm.2008.0006.

LaFollette, Marcel Chotkowski. *Science on American Television: A History*. University of Chicago Press, 2013.

———. *Science on the Air: Popularizers and Personalities on Radio and Early Television*. University of Chicago Press, 2009.

Lightman, Bernard. *Victorian Popularizers of Science: Designing Nature for New Audiences*. University of Chicago Press, 2009.

Males, Jennifer, and Peter Van Aelst. 'Did the Blue Planet Set the Agenda for Plastic Pollution? An Explorative Study on the Influence of a Documentary on the Public, Media and Political Agendas'. *Environmental Communication* 15:1 (2021): 40–54. https://doi.org/10.1080/17524032.2020.1780458.

Mason-Wilkes, Will. 'Divine Dna? "Secular" and "Religious" Representations of Science in Nonfiction Science Television Programs'. *Zygon* 55:1 (2020): 6–26. https://doi.org/10.1111/zygo.12574.

———. 'Science as Religion? Science Communication and Elective Modernism'. Phd, Cardiff University, 2018. http://orca.cf.ac.uk/109735/.

Matthes, Jörg, and Matthias Kohring. 'The Content Analysis of Media Frames: Toward Improving Reliability and Validity'. *Journal of Communication* 58:2 (2008): 258–79. https://doi.org/10.1111/j.1460-2466.2008.00384.x.

Milam, Erika Lorraine. *Creatures of Cain: The Hunt for Human Nature in Cold War America*. Princeton University Press, 2019.

Miller, David. *The Circuit of Mass Communication: Media Strategies, Representation and Audience Reception in the AIDS Crisis*. SAGE, 1998.

Sacks, Jonathan. "Without God we tend to start worshipping ourselves", *The Times*, February 23, 2013, 79

Scannell, Paddy, and David Cardiff. *A Social History of British Broadcasting: Volume 1 – 1922–1939, Serving the Nation*. Wiley, 1991.

Seaton, Jean. *Pinkoes and Traitors: The BBC and the Nation, 1974–1987*. Profile Books, 2015.

Secord, James A. *Victorian Sensation: The Extraordinary Publication, Reception, and Secret Authorship of Vestiges of the Natural History of Creation*. University of Chicago Press, 2003.

Sendall, Bernard. *Independent Television in Britain: Origin and Foundation 1946–62*. Springer, 1982.

Silverstone, Roger. *Framing Science: The Making of a BBC Documentary*. BFI Publishing, 1985.

Smocovitis, Vassiliki Betty. *Unifying Biology: The Evolutionary Synthesis and Evolutionary Biology*. Princeton University Press, 1996.

Telotte, J. P. *The Essential Science Fiction Television Reader*. University Press of Kentucky, 2008.

"The Wonder Years", *The Times*, September 1, 2011, 2.

Turney, Jon. 'Telling The Facts Of Life: Cosmology and the Epic of Evolution'. *Science as Culture* 10:2 (2001): 225–47. https://doi.org/10.1080/09505430120052301.

Unsworth, Amy, and David Voas. 'The Dawkins Effect? Celebrity Scientists, (Non) Religious Publics and Changed Attitudes to Evolution'. *Public Understanding of Science* (2021). https://doi.org/10.1177/0963662521989513.

Wilson, Edward O. *On Human Nature*. Harvard University Press, 1978.

Wulf, Andrea. *The Invention of Nature: Alexander Von Humboldt's New World*. Alfred A. Knopf, 2015.

Zakariya, Nasser. *A Final Story: Science, Myth, and Beginnings*. University of Chicago Press, 2017.

CHAPTER 2

Situating the Story: The Early Years of Evolution on the Wireless

The Stream of Life

Picture the scene; it's a September evening in 1925 in a small town in the north of England, the kids are long in bed, you've read the evening newspaper and you flick on your new wireless set, a recent purchase, after the launch the previous year of the Leeds/Bradford relay of the radio station BBC 2ZY, broadcasting from nearby Manchester. After some tweaking, over a faint crackling signal you hear an authoritative, but calm and clear male voice:

> The discoveries of biological science during the last hundred years have quite altered our outlook upon many familiar aspects of human life. One of the spheres where the change has been most obvious is in regard to heredity and racial problems. In the old days very little attention was given to such topics; but to-day a new spirit is abroad. Man now realizes that knowledge may make him master of his own fate, and of the future of his race, in ways undreamt-of by his ancestors. Blind acquiescence in destiny is giving place to the hope that destiny may in large measure come to be controlled.[1]

[1] Julian Huxley, *The Stream of Life* (London: Watts, 1926), 1. Excerpt from *The Stream of Life* by Julian Huxley reprinted by permission of Peters Fraser & Dunlop (www.petersfraser-dunlop.com) on behalf of the Estate of Julian Huxley.

A. Hall, *Evolution on British Television and Radio*, Palgrave Studies in Science and Popular Culture,
https://doi.org/10.1007/978-3-030-83043-4_2

Fig. 2.1 Julian Huxley recording an early BBC radio broadcast, circa 1930. BBC © BBC Photo Library

And thus began *The Stream of Life*, the BBC's first nationally broadcast content on biological evolution. One of the largest forays into science broadcasting to date for the fledgling national radio service, the 15-minute talk was the first in a series of eight on evolution, by the biologist and well-known public figure Julian Huxley (Fig. 2.1).[2] The fortnightly series was broadcast to a then relatively small—only 2.3 million radio licences had been issued by 1927—but disproportionally metropolitan elite audience.[3] The lectures, delivered in the dry and didactic style common across BBC content of the period, gave the listener a cursory history of evolutionary thought, a basic understanding of the latest developments in the field and

[2] Founded in 1922 as the British Broadcasting Company Ltd, following the recommendations of the Crawford Committee, the British Broadcasting Corporation, a non-commercial Crown Chartered organisation replaced the original consortium of private companies, on January 1, 1927. For more on the institutional history of the BBC in this period, see Asa Briggs, *The History of Broadcasting in the United Kingdom: Volume I: The Birth of Broadcasting* (Oxford University Press, 1961).

[3] While broadcast coverage was increasing year on year during this period, the BBC's radio service was primarily broadcast from regional stations in major urban centres via their associated transmitters. In the case of *The Stream of Life* although made in London the broadcast went out on the following stations: Aberdeen – 2BD, Belfast – 2BE, Birmingham – 5IT, Bournemouth – 6BM, Cardiff – 5WA, Glasgow – 5SC, London – 2LO, Manchester – 2ZY, Newcastle – 5WO. BBC Genome Project, http://genome.ch.bbc.co.uk and BBC, *The British Broadcasting Corporation Handbook*, (HMSO, 1928), 161–184.

Table 2.1 Episodes of *The Stream of Life*

	Episode title	*Date broadcast*
1	The Continuity of Life	21 Sept 1925
2	Inheritance[a]	5 Oct 1925
3	Development	19 Oct 1925
4	Evolution: I. The Problem	2 Nov 1925
5	Evolution: II. The Solution	16 Nov 1925
6	Nature and Nurture	30 Nov 1925
7	The Evolution of Man	14 Dec 1925
8	The Hope of Betterment	28 Dec 1925

BBC Genome Project, http://genome.ch.bbc.co.uk

[a]In the book version of the lectures this episode was renamed "Heredity". Huxley, *The Stream of Life*, 6

a sense of the importance of this field of scientific enquiry for society and humanity itself. Largely typical of Huxley's wider writings on evolution in the period, the series covered eight areas (Table 2.1) and after "numerous requests" was published almost verbatim as a book the following year.[4]

Although only 38 years old in 1925, Julian Huxley to date had a remarkable, but somewhat probable, career path. The grandson of the biologist Thomas Henry Huxley (1825–1895)—often known as "Darwin's Bulldog"—Julian Huxley had graduated from Baliol College, Oxford, in 1909 with a first-class honours degree in natural science. At just 26, he had launched a new biology department at Rice University in the US, and when he broadcast *The Stream of Life* in 1925 had recently taken up a chair in Zoology at King's College, London.[5] While Huxley would continue to identify as an evolutionary biologist and remained respected by world-leading scientists in the discipline, not long after these broadcasts in 1927 he gave up his Professorship at King's College, and would never hold another salaried academic post again. Instead, he dedicated the rest of his career to a multitude of activities centred on communicating science to popular audiences.[6]

[4]Huxley, *The Stream of Life*, Preface. For an overview of Huxley's popular science writing during this period, see Peter J. Bowler, *Science for All: The Popularization of Science in Early Twentieth-Century Britain* (University of Chicago Press, 2009), 221–227.

[5]Robert Olby, "Huxley, Sir Julian Sorell," *Oxford Dictionary of National Biography*, 2004, https://doi.org/10.1093/ref:odnb/31271.

[6]For more on Huxley's career, see C. Kenneth Waters, *Julian Huxley, Biologist and Statesman of Science: Proceedings of a Conference Held at Rice University, 25–27 September 1987* (Rice University Press, 1992).

While *The Stream of Life* was largely "Darwinian" in the outline of evolutionary theory sketched by Huxley, the series did introduce listeners to several newer areas of biological research, which by the 1920s were seriously challenging the mechanisms of evolution as originally proposed by Charles Darwin. In particular in the second episode, by utilising simple mathematical examples detailing the role of recessive genes in patterns of inheritance, Huxley introduced the audience to Mendelian inheritance. Huxley's series discussed the latest biological understandings of evolution and was in stark contrast to other popular science media on the subject in the period. Take, for example, the second edition of clergyman and educator Dennis Hird's *A Picture Book of Evolution* published in 1920, which used alluring but erroneous examples of the technological development of human-invented machinery, such as the bicycle, to illustrate examples of biological evolution.[7] Despite criticism, *A Picture Book of Evolution*, with its attempt to place biological evolution within an all-encompassing process of universal evolution, remained popular into the late 1940s.[8]

A prominent scientist like Huxley presenting such cutting-edge evolutionary research to a general audience, via the new medium of wireless radio, was an extremely important milestone. The contemporary scientific research featured in the broadcast, in particular that on heredity, would go on to become central to the modern evolutionary synthesis—the fusion of Darwinian evolution with Mendelian genetics, which during the 1930s and 1940s was purported by its proponents, including Huxley, to have unified the disparate fields of biological study.[9] The emergence and popularisation of the modern synthesis was a slow, protracted and contested process. The first major work that attempted to combine Mendelian genetics with natural selection, Ronald Fisher's *The Genetical Theory of*

[7] The use of human technology to illustrate the process of biological evolution had a long tradition in print. For example see Samuel Butler, "Darwin among the Machines," *The Press*, Christchurch, New Zealand, June 13, 1863. Available in full online http://www.gutenberg.org/files/6173/6173-h/6173-h.htm *(last accessed 05/06/2021)*.

[8] *A Picture Book of Evolution* was advertised in elite publications, such as *The Freethinker* (e.g. "Watts & Co's Publications", *The Freethinker*, February 22, 1925, Vol. XLV, No. 8, 127). For an example of a review criticising the veracity of the content of the book see: "Books of the Day", *The Manchester Guardian*, October 11, 1932, 5. The 4th edition of the book, updated and amended by C. M. Beadnell, was published in 1948 and can be viewed online: https://archive.org/details/in.ernet.dli.2015.18999 *(last accessed 17/06/2021)*.

[9] For more on the modern evolutionary synthesis see: Vassiliki Betty Smocovitis, *Unifying Biology: The Evolutionary Synthesis and Evolutionary Biology* (Princeton University Press, 1996). In particular chapter six, "Reproblematizing the Evolutionary Synthesis", 191–210.

Natural Selection, was not published until 1930, and it was not until 1942 that Huxley himself coined the phrase the "modern synthesis".[10] While *The Stream of Life* was by no means a rounded-out synthetic vision of biological unification, beyond its scientific content it also contained several other elements that became features of the modern synthesis, or at least Huxley's version of it. Infused with Huxley's "evolutionary humanism", these broadcasts from 1925, several years prior to many of the scientific discoveries central to the modern synthesis,[11] add further support to the argument that the construction of this unified vision of biology, particularly in its British form, was as much a philosophical and socio-political endeavour, as it was based upon pure scientific knowledge.[12]

Although careful to limit language that implied there was a directional or purposeful aspect to evolutionary development—known as teleology—the main narrative thread of Huxley's script in *The Stream of Life* was centred on progress; progress in scientific knowledge. Huxley made a clear distinction between other species and *Homo sapiens*, who, he argued, through our own self-consciousness and accumulated knowledge, could now control our own evolution, and thus destiny.[13]

Despite its cutting-edge content and utilisation of a progressive narrative, which add to *The Stream of Life's* importance in the history of science broadcasting, we have little direct evidence of the audience's reception of the broadcast and accompanying book. While the show was listed in the radio listings of most national daily newspapers, it seems to have elicited little response or comment from journalists.[14] As already introduced in

[10] Julian Huxley, *Evolution, the Modern Synthesis* (Allen & Unwin, 1942).

[11] For example, by 1925 only the first two of J. B. S. Haldane's influential ten paper series, *A Mathematical Theory of Natural and Artificial Selection* had been published.

[12] See: Smocovitis, *Unifying Biology*; Maurizio Esposito, 'Utopianism in the British Evolutionary Synthesis', *Studies in History and Philosophy of Science Part C: Studies in History and Philosophy of Biological and Biomedical Sciences*, Defining Darwinism: One Hundred and Fifty Years of Debate, 42:1 (2011): 40–49, https://doi.org/10.1016/j.shpsc.2010.11.007; and Paul Weindling, 'Julian Huxley and the Continuity of Eugenics in Twentieth-Century Britain', *Journal of Modern European History* 10:4 (2012): 480–499, https://doi.org/10.17104/1611-8944_2012_4.

[13] Huxley presented a strikingly similar, albeit more expansive, version of this vision in the preface to his collected *Essays of a Biologist* published just two years earlier in 1923. Julian Huxley, *Essays of a Biologist* (A. A. Knopf, 1923), available online in full: http://archive.org/details/essaysofbiologis1923huxl *(last accessed 17/06/2021)*.

[14] For an example of radio listing see "Some Coming Wireless Talks: The King's Speech," *The Manchester Guardian*, September 26, 1925, 13.

Chap. 1, and as we shall see repeatedly throughout this book, broadcast media is by its very nature ephemeral, often leaving a limited historical footprint. However, a lack of reviews or audience feedback for a specific show does not mean we cannot begin to understand its impact and potential influence as one component of a wider tapestry of science media on a specific subject. In addition to the potential cumulative effects of the progressive narrative and subject framing first transferred to radio by Huxley, we can also consider the indirect influences that the series may have had. While *The Stream of Life* was new content commissioned by the BBC, it overlapped closely with Huxley's other popular outputs during the period.[15] The radio broadcasts are also in-keeping with Huxley's later work, particularly his time as Secretary of the Zoological Society of London, where he sought to promote a "general biology", which centred on explanatory, analytical and synthetic aspects of the subject.[16] The show's biggest influence may well have come from a more intangible process of Huxley's ideas being adapted and disseminated via other popular writers, broadcasters and social commentators: rippling outward from a small, educated and science-minded metropolitan elite who initially tuned in to the broadcasts, and who may have also read Huxley's other popular works in the period. For example, in the 1930s, journalist and former editor of *The Eugenics Review*, Eldon Moore, published a chapter on heredity in the biology focused volume of a popular home encyclopaedia, *The Standard University* with the title, "Heredity: The Stream of Life".[17] Moore's chapter presents the standard accepted science on heredity as it stood at the time, and perpetuated the connection of evolution with the phrase, "the stream of life", as first made by Julian Huxley for his BBC series in 1925. Moore and Huxley were peers, but it is unclear whether Huxley influenced Moore's chapter via the original broadcasts, the subsequent book or through personal communication.[18] In this instance we may only be

[15] See, for example, Huxley, *Essays of a Biologist*.

[16] Joe Cain, 'Julian Huxley, General Biology and the London Zoo, 1935–42', *Notes and Records of the Royal Society* 64:4 (2010): 359–378, https://doi.org/10.1098/rsnr.2010.0067.

[17] J.M. Parrish and John R. Crossland (Eds), *The Science of Living Things*, The Standard University (Odhams Press Ltd, 1935), 455–488. For a short obituary of Eldon Moore see: 'Mr. Eldon Moore.' *The Times*, November 22, 1954, 8.

[18] Historical records show Huxley and Moore's paths crossing upon several occasions, largely through Moore's role as editor of *The Eugenics Review* and other journalistic endeavours. For example, Moore referenced Huxley's contributions in a report from the first World

tracing the transmission of a rather innocuous phrase, but this example alerts us to the usually hidden and indirect ways that ideas can be communicated across different forms of popular media in society. For while it is often impossible to meaningfully assess the influence of one broadcast or series, we can begin to explore how more generalised theses or ideas, and the dominant narrative framings that accompany them, are reproduced, transmitted and popularised via cumulative effects across different mediums.

While the early history of radio broadcasting in the UK is outlined in some detail by general histories of British broadcasting, to date no comprehensive history of science broadcasting on the radio exists.[19] Where studies of science broadcasting in twentieth-century Britain do exist, they overwhelmingly focus on television,[20] or overarching top-level BBC policy.[21] There are, however, a few relatively recent endeavours, which outline in detail specific, contingent episodes in the early history of BBC science broadcasts on the radio. In particular the work of historian Allan Jones highlights that Huxley's first broadcasts as introduced in this chapter were part of a wider emergent, largely unregulated and unplanned, landscape of science broadcasting at the BBC. These early science broadcasts in the 1920s were driven by a few science communicators and science-minded producers at the BBC who were interested in ensuring science content

Population Conference in Geneva in 1927, where both were in attendance, see: Eldon Moore, "The World Population Conference," *The Spectator*, September 10, 1927, 8.

[19] For a general introduction to the history of radio broadcasting in the UK see Asa Briggs, *The History of Broadcasting in the United Kingdom: Volume II: The Golden Age of Wireless* (Oxford University Press, 1965); and Sean Street, *Historical Dictionary of British Radio* (Rowman & Littlefield, 2006).

[20] For example, Jean-Baptiste Gouyon, 'From Engaged Citizen to Lone Hero: Nobel Prize Laureates on British Television, 1962–2004', *Public Understanding of Science* 27:4 (2018): 446–457, https://doi.org/10.1177/0963662518760790; Timothy Boon, '"The Televising of Science Is a Process of Television": Establishing Horizon, 1962–1967', *The British Journal for the History of Science* 48:1 (2015): 87–121, https://doi.org/10.1017/S0007087414000405; and Rupert Cole, '1972: The BBC's Controversy and the Politics of Audience Participation', *Public Understanding of Science* 26:4 (2017): 514–518, https://doi.org/10.1177/0963662516684231.

[21] See, for example, Timothy Boon, *Films of Fact: A History of Science in Documentary Films and Television* (Wallflower Press, 2008); and Allan Jones, 'Elite Science and the BBC: A 1950s Contest of Ownership', *The British Journal for the History of Science* 47:4 (2014): 701–723, https://doi.org/10.1017/S0007087413000927.

became a permanent fixture of the new medium.[22] Through these early efforts, science talks and lectures settled into the regular schedule, with both regionally and nationally produced content on subjects as diverse as "Medicinal Plants and Their Uses" and "Eclipses of the Sun" appearing during the second half of the 1920s.[23] These early broadcasts, especially those created in-house at the regional stations, often featured local academics, science writers or journalists, and other enthusiasts of the subject at hand. Yet as radio's popularity continued to grow, and the BBC's reach from the capital became nationwide,[24] the spaces for speaking about science on the radio quickly became monopolised by a small number of academics and science popularisers associated with elite institutions and societies.[25]

Following *The Stream of Life*, the BBC broadcast a steady number of talks about biological evolution, featuring other prominent academics such as Professor Grafton Elliott Smith, Chair of Anatomy at University College London and Professor of Botany at the University of Cambridge, Albert Seward.[26] Of the 39 broadcasts that featured the word "evolution" in their advertised listing during the period 1925–1944 (Fig 1.1), all but one were simple scripted lectures aimed at either adult or school-level audiences. This format, which replicated earlier traditions of public lectures and popular single-authored monographs, was limited in its ability to interest new audiences, and educate those who didn't already have an

[22] Allan Jones, 'Mary Adams and the Producer's Role in Early BBC Science Broadcasts', *Public Understanding of Science* 21:8 (2012): 968–983, https://doi.org/10.1177/0963662511419450.

[23] *Mr A. O. Bentley Ph.C*, "Medicinal Plants and Their Uses". BBC Nottingham – 5NG, 9 March 1925; and *Prof. H. H. Turner – I*, "Eclipses of the Sun". Written and presented by Professor H. H. Turner. BBC London – 2LO, 24 May 1927.

[24] The BBC began using longwave transmission from a station in Daventry (5XX) in 1925, and in March 1939 this transmitter began broadcasting, the *National Programme*, a service programmed centrally in London and broadcast nationwide. For more see Briggs, *The History of Broadcasting in the United Kingdom*, Chapter V.

[25] For early examples of learned societies contacting the BBC with suggestions for science content see the file *Talks. Science, File 1B, 1937–1941*, R51/523/2, BBC Written Archives, Reading, UK (BBC-WA).

[26] See *Prof. G. Elliott Smith*, "The Movements of Living Creatures". Written and presented by Grafton Elliott Smith. BBC London – 2LO, January 17, 1927; and *The Origins of Life – III*, "The Evolution of Plants and the Formation of Coal". Written and presented by Albert Seward. BBC London – 2LO, November 19, 1929.

interest in the subject being discussed.[27] Although during these early years of radio, senior figures at the BBC reflected on the great educational and egalitarian potential of the medium, few gave detailed consideration to the specific needs of science broadcasts.[28] This problem was partly exacerbated by the fact that there was no department, or team within a department at the BBC dedicated solely to science broadcasts.

Controlling the Content, While Developing the Format(s)

Most science content produced during the first two decades at the BBC was created by the Schools or Adult Education sections of the Education department.[29] Of the producers involved with these early endeavours, the most notable was the pioneering Mary Adams (1898–1984), who after being a successful guest lecturer on a series of six talks in 1928 on the *Problems of Heredity*, joined the Adult Education team in 1930.[30] A botany graduate from University College Cardiff, prior to her appointment, Adams had been an education tutor and was already a member of the British Institute of Adult Education.[31] Before joining the BBC, Adams had already begun thinking about and reflecting upon the specific needs and potential of science broadcasting, as summarised by the historian Allan Jones:

> She considered that, in a broadcast talk, ideas counted for more than facts, which often had to go unstated. A good voice was more important than subtlety of expression. Finally, "[b]roadcast talks need not be confused with

[27] For more on popular science print formats, their development and publisher's (lack of) ability to reach new audiences, see Bowler, *Science for All*, 75–113.

[28] See Reith, J. *Broadcast over Britain* (Hodder and Stoughton, 1924), and Matheson, H. *Broadcasting* (Butterworth, 1933), as introduced and contextualised in Allan Jones, 'Mary Adams and the Producer's Role in Early BBC Science Broadcasts', *Public Understanding of Science* 21:8 (2012): 969–972, https://doi.org/10.1177/0963662511419450.

[29] Both Schools and Adult Education became their own departments outright during the 1930s; see Chap. 4.

[30] Jones, 968. For more on Adams' career at the BBC see Kate Murphy, *Behind the Wireless: A History of Early Women at the BBC* (Springer, 2016), 181–186.

[31] Murphy, *Behind the Wireless*; and Sally Adams, "Adams, Mary Grace Agnes (1898–1984)," *Oxford Dictionary of National Biography*, Oxford University Press, 2004, http://www.oxforddnb.com/view/article/30750.

education: they are promising stimulants which may become a powerful aid to the spread of education". In other words, mainstream broadcasting (as opposed to specialist broadcasting such as Schools broadcasts) was not a medium for delivering lectures; but, used in the right way, it could surprise casual listeners into wanting to know more.[32]

Adams set about producing science broadcasts that emphasised these qualities, favouring discussion of the process of science over pure scientific knowledge and prioritising clear and concise delivery over scientific seniority. One result of this approach was an increased reliance on science broadcasters to present regular magazine content, most notably the writer Gerald Heard (1889–1971) who began acting as a regular science correspondent starting with the fortnightly *Research and Discovery* in the summer of 1930.[33] Heard, much like his contemporary Julian Huxley, had rejected doctrinal Christianity in favour of humanist philosophies while studying at university. However, as his second book *The Ascent of Humanity* (1929), which explores the evolution of the body-mind of man with regard to human progress towards a state of super-consciousness shows, Heard remained keenly interested in spiritual matters throughout his varied and ranging career.[34]

The strong vision and leadership of the BBC's first Director-General, John Reith (1889–1971) influenced the emergence of a close and paternalistic style of broadcasting management at the BBC. So much so that today, the byword *Reithian* is still used to encapsulate his puritanical approach and mission to "inform, educate and entertain".[35] Although

[32] Jones, 968–969.

[33] "Research and Discovery". Presented by Gerald Heard. BBC National Programme, May–August 1930.

[34] Alison Falby, "Heard, Henry Fitzgerald [Gerald] (1889–1971)," *Oxford Dictionary of National Biography*, Oxford University Press, 2013, https://doi.org/10.1093/ref:odnb/54040; Gerald Heard, *The Ascent of Humanity: An Essay on the Evolution of Civilization from Group Consciousness Through Individuality to Super-Consciousness* (Harcourt, Brace, 1929).

[35] Reith actually borrowed the phrase, originally "entertaining, educating and informing" from the American broadcaster David Sarnoff who first used it in 1922. The slightly reformulated version to "inform, educate and entertain" is still at the heart of the BBC's mission statement. "This is the BBC 1922–1926," *History of the BBC*, https://www.bbc.co.uk/historyofthebbc/research/culture/reith-5, *(last accessed 02/04/2019)*; and "Broadcasting: Copy of Royal Charter for the continuance of the British Broadcasting Corporation", Cm 9365, December 2016. http://downloads.bbc.co.uk/bbctrust/assets/files/pdf/about/how_we_govern/2016/charter.pdf, *(last accessed 17/06/2021)*. For more on Reith's early

early departmental organisation was largely structured around mode of production rather than broadcast subject,[36] many subject areas received dedicated and controlled oversight from senior figures at both the departmental and executive levels. In addition, unlike the disparate and largely ad-hoc early science broadcasts, content in other areas was overseen and regularly reviewed by a series of independent advisory committees. For example, from 1925 onwards the Central Religious Advisory Committee (CRAC) not only oversaw broadcast content created by the Religion Department, but also regularly reviewed a wide range of broadcasts deemed to have relevant ethical, moral or philosophical content.[37]

As in wider British society during the period, the BBC used the word "religion" synonymously with Christianity. CRAC considered religious broadcasting's mission to be "evangelical" and when the BBC received solicitations from members of Christian denominations who were not being represented, they advised the BBC to clarify that output was only for those "bodies which are in the main stream of historic Christianity" within the UK.[38] As well as Christian denominations excluded under this definition, such as Unitarians, the BBC also excluded rationalist and humanist organisations, and any other anti- or non-Christian worldviews.[39] As outlined by the historian of science Robert Bud, this opposition to humanism being allowed proportionate representation alongside religious programming at the BBC resulted in its emergence across other areas of BBC programming.[40] Aspects of humanist worldviews popped up on

vision for the organisation see: John Reith, *Broadcast over Britain* (Hodder and Stoughton, 1924).

[36] The exception to this pattern was the Music department.

[37] See BBC-WA file R6/14, "Central Religious Advisory Committee, Correspondence, 1925–1964". For an overview of the early development of religious broadcasting at the BBC see: Francis House, "Review of the Aims and Achievements of Religious Broadcasting, 1923–1948", BBC paper presented to CRAC, 14 September 1948, R6/22/1, BBC-WA.

[38] Francis House, "Review of the Aims and Achievements of Religious Broadcasting, 1923–1948", BBC paper presented to CRAC, 14 September 1948, R6/22/1, BBC-WA; and Kathleen Bliss, Leslie E. Cooke, T.W. Manson, and J. Masterson, "The Main Stream of Historic Christianity", Sub-committee paper presented to CRAC, R6/22/1, BBC-WA. All BBC copyright content reproduced courtesy of the British Broadcasting Corporation. All rights reserved.

[39] "Humanism and Religious Broadcasting," BBC Paper presented to CRAC, 7 October 1958, R6/21/8, BBC-WA.

[40] Robert Bud, '"The Spark Gap Is Mightier than the Pen": The Promotion of an Ideology of Science in the Early 1930s', *Journal of Political Ideologies*, 22:2 (2017), https://doi.org/10.1080/13569317.2017.1298548.

shows made by departments as diverse as Radio Talks and Schools Education yet they were almost always contrasted against, or profiled alongside, more traditional Christian beliefs.[41] For example, in 1943, when Julian Huxley pitched a radio talk about humanism, the BBC gave him the go-ahead only on the premise that the talk was one of a series of three, with his talk on "Scientific Humanism" being followed by talks on "Classical Humanism" and "Christian Humanism".[42]

One area where humanist views managed to avoid this caveated and controlled framing was in scientific programming, where, as outlined by Bud, from the 1930s onwards a form of "scientific humanism" came to dominate the BBC's scientific output.[43] Spearheaded by Huxley and predominantly produced by Adams,

> [c]ollectively these widely attended-to talks defined a series of related discourses which together promoted the interestingness of science. They went beyond this, however, to deal with the major issues of religion, civilization and the nature of knowledge. In this way they could be seen as using talk about science as a proxy for scientific humanism.[44]

Many of these broadcasts were presented by Adams' new go-to science broadcaster, Gerald Heard, who had worked with Julian Huxley as the Literary Editor of the short-lived "Journal of Scientific Humanism", *The Realist* from 1929 to 1930.[45] As well as receiving much larger audiences than the ill-fated humanist magazine, following the example set by Huxley's *Stream of Life*, many of these early science broadcasts were also accompanied by popular books.[46] Alongside their broadcast efforts, this small group of Huxley's rationalist peers, which included the evolutionary

[41] For examples from Radio Talks and Schools Education see respectively: *Key Words.* "Christians and Humanists." BBC Home Service, February 8, 1956; and *For Schools and Colleges.* "Living in the Present: Belief." BBC One, November 29, 1965.

[42] See "Rcont1 Talks. Sir Julian Huxley, File II, 1943–1949," BBC-WA.

[43] For more on the ideological depiction of science in television drama, see Robert G. Dunn, 'Science, Technology and Bureaucratic Domination: Television and the Ideology of Scientism', *Media, Culture & Society* 1:4 (1979): 343–354, https://doi.org/10.1177/016344377900100403.

[44] Bud, '"The Spark Gap Is Mightier than the Pen"', 7.

[45] Bud, 5–6.

[46] See, for example, Gerald Heard, *Science in the Making* (London: Faber and Faber, 1935) which was based on the BBC radio series of the same name which ran in various guises from 1931 to 1935.

biologist J. B. S. Haldane (1892–1964), also published other popular books connecting a wide range of subjects to their incipient conception of a progressive scientific humanism. In the opening edition of *The Realist*, the historian of science Charles Singer remarked on the "lamentable separation of humanistic from scientific studies", calling for understandings of science to go beyond the knowledge-focused "direct" study of science in the classroom, to incorporate the much more influential "indirect" effects on "men's thinking" across all aspects of society. In his essay *Scientific Humanism*, Singer argued that if:

> [T]he sole plea for science in education would be its practical application ... science would need to be numbered among the crafts and tricks of life. It would be of purely vocational value. This is an unworthy and, I believe, untrue view of science which would rob it of all spiritual value.[47]

Reflecting Huxley's views on the matter, the magazine remained respectful of religion; "we believe that if it can be understood it need not be repressed either by neglect, ridicule or force, but that it can and should have worthy expression". Yet, like any "social movement" in regard to religion, the magazine's editors aimed to "analyse it and attempt to understand and explain it".[48] Thus Huxley and his peers on the editorial board of *The Realist* saw scientific humanism as the next natural step in man's cultural and ultimately biological evolution.[49] As the historian of science Vassiliki Betty Smocovitis states: "[f]or Huxley 'scientific humanism', later more popularly known as 'evolutionary humanism', was more than just an abstraction of a philosophical exercise; it was instead a central feature of his world view, his philosophy of life and simultaneously of his scientific and political endeavours."[50]

47 Charles Singer, "Scientific Humanism", *The Realist*, 1:1, April 1929, 12–18.

48 Archibald Church, "Editorial", *The Realist*, 1:1, April 1929, 180.

49 While originally conceived and most commonly promoted by Julian Huxley, scientific humanism was also forwarded by others beyond his immediate social circles. For example see the US philosopher, Oliver Reiser, *The promise of scientific* humanism *toward a unification of scientific, religious, social and economic thought*, (University of Wisconsin, 1940). For more on the historian of science George Sarton's views on scientific humanism and the integral role the history of science could play in promoting this secular worldview, see Robert Westman, 'The "Two-Cultures" Question and the Historiography of Science in the Early Decades of the Salk Institute for Biological Studies', *Sartoniana* 32 (2019), 58–60.

50 Vassiliki Betty Smocovitis, "The unifying vision: Julian Huxley, evolutionary humanism and the evolutionary synthesis" in Harmke Kamminga and Geert Somsen, *Pursuing the*

However, as the short shelf-life of *The Realist* attests, we must remember that the thinkers and elites loosely associated around Huxley did not share a single vision or conception of scientific humanism, or the other various forms of progressive humanism, which were emerging during the inter-war years. For example, as previously introduced, in 1929 Heard published, *The Ascent of Humanity: An Essay on the Evolution of Civilization from Group Consciousness Through Individuality to Super-Consciousness*, which although evidently anti-Christian, maintained a role for the unknown and supernatural via Heard's focus on the evolution of man's spirit.[51] Further, as Bud demonstrates, while science radio broadcasts during the 1930s that featured scientific humanism were part of the wider "Social Relations of Science" movement, we should not lose sight of the lack of coherence, fluidity and interchanging terminology used by many of the concept's proponents.[52]

As the 1930s progressed, different ways of presenting science on the radio continued to be trialled, developed and further refined. Reflecting a wider movement that aimed to promote the utility of science for society, programmes that spoke about science in society, rather than as an abstract entity separate from everyday life, proved a popular way of presenting radio science features.[53] In addition to this approach of talking about science and society, producers within the Features department also saw the utility of developing content that informed the listener about the process of science and the practice of scientists themselves, rather than just the regurgitation of dry scientific facts, theories or discoveries.[54]

Before leaving to join the nascent BBC television project in 1936, Mary Adams presided over at least 20 radio broadcasts about evolution, including several in the science and society mould, such as *Biology in the Service of Man* (1932) written and presented by the naturalist, popular science

Unity of Science: Ideology and Scientific Practice from the Great War to the Cold War (Routledge, 2016), 33.

[51] Gerald Heard, *The Ascent of Humanity: An Essay on the Evolution of Civilization from Group Consciousness Through Individuality to Super-Consciousness* (Harcourt, Brace, 1929).

[52] Bud, '"The Spark Gap Is Mightier than the Pen"', 2–4.

[53] Jones, 'Mary Adams and the Producer's Role in Early BBC Science Broadcasts', 977–979.

[54] A prime example is the 1936 series *Scientists At Work*; for more see "Talks. Science, File 1A, 1936", R51/523/1, BBC-WA.

writer and noted neo-vitalist J. Arthur Thomson (1861–1933).[55] Adams' unwavering belief in the power of education, radio's ability to educate new audiences, and the importance of broadcasting of evolutionary science to the masses were all influenced by her eugenicist beliefs. Similar to Huxley's views on race, class and education as alluded to in *The Stream of Life*,[56] Adams believed that as the working classes reproduce at a higher rate than the rest of society, universal education was required to ensure their offspring were less "empty-headed, indolent or idle", than their parents.[57] While such views may seem jarring to the modern reader, eugenic ideas such as this were commonplace among British intellectuals during the period.[58]

Julian Huxley was at the centre of many of these early broadcasts, if not as a writer and presenter, acting as a gatekeeper, directing the BBC towards other suitable academics. He was one of the first scientists to realise the potential of the new mass mediums of radio, film and later television.[59] Further, as different ways of speaking about science on the radio coalesced, this small group of public intellectuals tasked with communicating complex scientific theories to the public increasingly found they could also act

[55] Vitalists and neo-vitalists believed that there is a fundamental non-physical element, a "vital spark" or "élan vital", which separates organisms from non-living entities. "Biology in the Service of Man." Written and Presented by J. Arthur Thomson, produced by Mary Adams. BBC National Programme, May–July 1932. For more on J. Arthur Thomson see Peter J. Bowler, "From Science to the Popularisation of Science: The Career of J. Arthur Thomson", in David M. Knight and Matthew D. Eddy, *Science and Beliefs: From Natural Philosophy to Natural Science, 1700–1900* (Routledge, 2017) 231–250; and Bowler, *Science for All*, 233–240.

[56] Huxley, *The Stream of Life*, 50–56.

[57] Jones, 'Mary Adams and the Producer's Role in Early BBC Science Broadcasts', 11.

[58] For example, the editor of *The Realist* Archibald Church, who was also an MP during the period, in 1931 attempted to pass a Private Member's Bill through Parliament that would have seen the introduction of sterilisation for those with mental defects.

[59] Huxley was also involved with many projects in these areas beyond his work with the BBC; for example, having produced an Oscar-winning and pioneering natural history documentary, *The Secret Life of Gannets* released in 1934 (watch online: https://player.bfi.org.uk/free/film/watch-private-life-of-the-gannets-1934-online), and acting as an adviser to the Associated Realist Film Producers organisation (ARFP) founded by the documentary filmmaker Paul Rotha from 1936. See Timothy Boon, *Films of Fact: A History of Science in Documentary Films and Television* (Wallflower Press, 2008) 83–87; and Timothy Boon, "'To formulate a better plan for living' Visual communication and scientific planning in Paul Rotha's films, 1935–1945" in Harmke Kamminga and Geert Somsen, *Pursuing the Unity of Science: Ideology and Scientific Practice from the Great War to the Cold War* (Routledge, 2016), 161.

as experts on other subjects being broadcast. Scientism—the idea that rational thought and the scientific method could be applied to political and social issues—abounded during the 1930s, and several scientific intellectuals benefitted from wide media attention.[60] After his first broadcast in 1925, Huxley, the archetypal public intellectual of the age, appeared on all manner of broadcasts speaking about subjects as diverse as telepathy, humanism and bird watching.[61] As well as using science broadcasts to speak about humanism, Huxley and other public intellectuals of the period used appearances on a diverse range of broadcasts to bypass other aspects of editorial and advisory committee censorship. For example, Huxley was one of the original experts on the panel show *The Brains Trust* which began in 1941; a show where members of the public submitted their questions for the panel to answer and debate.[62] While Huxley was the go-to panellist for any questions on biology, or science more generally, he regularly answered questions on all manner of subjects from social planning to the pernicious and dangerous nature of superstitious beliefs.[63]

The overreach of expertise of the specialists on *The Brains Trust* did not go unnoticed, becoming the subject of parody in a *Just William* children's story published in 1945, which featured a Professor Knowle and a music-hall spoof Professor Know-all.[64] In the story, the perennial prankster and protagonist, schoolboy William Brown managed to switch the esteemed Professor Knowle from the BBC's *Brains Trust* with a satirical version of

[60] Declan Fahy, "The Celebrity Scientists: A Collective Case Study", PhD Thesis (Dublin City University, 2010), 13–17.

[61] See respectively: "The Telepathy Experiment." Presented by Julian Huxley. BBC, London – 2LO, 19 September 1928; *Bird Watching and Bird Behaviour*, "The Pleasure of Bird Watching." Written and Presented by Julian Huxley. BBC National Programme, 2 May 1930; and *Humanism*, "Scientific Humanism." Written and Presented by Julian Huxley. BBC National Programme, 5 December 1943.

[62] Originally called *Any Questions?* and broadcast on the BBC Forces Radio service from January 1941, due to popular demand the show quickly began being repeated on the Home Service and was renamed *The Brains Trust* in 1942, running on the Home Service until 1949. In 1955 the format was resurrected for BBC Television where it ran until 1961.

[63] The interested reader can watch a clip of Huxley answering questions from an episode of *The Brains Trust* broadcast in 1945, at the British Pathé online archive: http://www.britishpathe.com/video/the-b-b-c-brains-trust-answering-any-questions-1/ *(last accessed 30/03/2021).*

[64] Richmal Crompton, *William and The Brains Trust* (Macmillan Children's Books, 1989). For a synopsis and more information on *Just William* see the website 'Just William's Year': http://justwilliamsyear.co.uk/william-and-the-brains-trust/william-and-the-brains-trust *(last accessed 30/03/2021).*

the professor, played by a local comedian performing for RAF servicemen in the village hall next door. The real Professor, who was speaking at a church hall recreation of the popular radio show because he knew the Vicar "well enough to make him break his rule of never speaking outside London except to university audiences", spent the whole story completely oblivious to the farce unfolding around him, lost in his own thoughts hypothesising about the "bucolic" rural people he encountered:

> The boy, the Professor decided, was not very intelligent. He replied shortly to all the Professor's questions and appeared to take no interest in any of the subjects of conversation introduced by the Professor. It suggested to the Professor a possible subject for future discussion. Is the Town Child more Mentally Alert than the Country Child? Certainly this country child did not seem mentally alert.

While written as children's fiction, in this short story we catch a glimpse of how certain contemporary audiences interpreted the austere, serious and paternalistic way that experts speaking via the BBC addressed their audiences during this period. Riffing on common tropes and stereotypes, in particular the differences between simple country-life and modern urban elites, the story paints Professor Knowle as vastly knowledgeable, but at the expense of any practical and social understanding of real life. Complete with its resplendent reproduction of the *Brains Trust* boffins' idiosyncrasies, such as the slow chin stroke, the sometimes pejorative tone reserved for the experts in this short story reminds us of the limited ability of early radio science broadcasts to overcome the rigid class boundaries prevalent in British society in the period.

By operating across the BBC, Huxley successfully navigated some of the challenges that came with being a public figure; he kept speaking about evolution separate from other radio and later TV work, where he more explicitly spoke about humanism and secularism. While simultaneously, through his science broadcasts he managed to infuse discussion of evolution with aspects of scientific materialism, progressivism and humanist concepts. This is something that didn't go completely unnoticed by viewers, the religious press and occasionally the BBC hierarchy themselves. For example, on one occasion when Huxley was asked to reflect on what biology can do in the service of man, for the final episode of a 1943 radio series called *Reshaping Man's Heritage*, he quickly skipped over the biological and moved onto the sociobiological and political application of the latest scientific discoveries. His draft script caused alarm amongst senior executives with one stating, "I am afraid, however, that as it stands the talk will open us to the charge of using a scientific series as a peg for political and social generalisations".[65] It was too late to pull the episode, so following some minor tweaks it was broadcast largely as Huxley intended. An audience research report on the series recorded an "Appreciation Index" for Huxley's talk of 91 out of 100, the highest awarded to date for a talk on *any* subject. However, the report did note that:

> The only criticism of his presentation of the subject came from those who disagreed with his scientific materialism and stressed the inadequacy of science, in the absence of moral criticism, to solve the problems of the future.[66]

While we may see Huxley's attempts to infuse broadcasts with scientific humanism as part of wider secularising trends of the period, it is also important to remember that often there was no clear daylight between scientific humanist visions of human progress, and those built on natural theological traditions. In his broadcasts during the inter-war and war years, while always placing man within his phylogenetic tree, Huxley usually extolled the distinct, exceptional biological and social characteristics of *Homo sapiens*, and always presented a hopeful and redemptive vision of humanity's future.[67] In Huxley's talk in the *Reshaping Man's Heritage* series, simply titled "The Future", he called for the application of scientific knowledge to societal challenges, arguing that "the next big forward step in human evolution will come through our control of social and economic problems".[68] As in 1925 in *The Stream of Life*, when Huxley introduced the socio-political into a series with a biological focus, here again, despite his specifically biological brief, he focused on his vision for the successful technocratic and scientific management of a global society. Coming the year after he published *Evolution: The Modern Synthesis* (1942), in

[65] C. V. Salmon, "Huxley's final talk in 'Reshaping Man's Heritage'". BBC Minute, 29 March 1943, R51/528/2, BBC-WA. All BBC copyright content reproduced courtesy of the British Broadcasting Corporation. All rights reserved.

[66] Listener Research Department, "A Listener Research Report: "Reshaping Man's Heritage"". BBC audience research report, 14 April 1943, R51/528/2, BBC-WA. All BBC copyright content reproduced courtesy of the British Broadcasting Corporation. All rights reserved.

[67] See for examples chapter VII – The Evolution of Man, and VIII – The Hope of Betterment in Huxley, *The Stream of Life*, and Julian Huxley, *The Uniqueness of Man*, (Chatto & Windus, 1943).

[68] J. S. Huxley et al., *Reshaping Man's Heritage: Biology in the Service of Man* (G. Allen & Unwin, 1944), 90–91.

Reshaping Man's Heritage we see stripped bare the utopianism and positivism driving Huxley and his close peers in the period. As the historian of science Maurizio Esposito has demonstrated, this progressive vision of humanity's future is inextricably bound with the British proponents of the evolutionary synthesis.[69] Even within broadcasts presented completely within a biological framework, such as *The Stream of Life*, Huxley's views on the potential for the future development of mankind are intimately linked to the specific socio-political climate of the period. In particular, the positivist visions put forth across the *Reshaping Man's Heritage* series from January to April 1943 were shaped by the conflict then raging around the world. In the eyes of the scientific elite who featured on this series, the war threatened to negatively co-opt the recent successes of scientific and technological developments in the public's conscience.[70] Huxley's talk was the final episode in a series of twelve, which had been opened by the writer H. G. Wells (1866–1946), and featured several eminent and distinguished scientists, including J. B. S. Haldane on *Reshaping Plants and Animals* and the zoologist James Watson Munro (1888–1968) and naturalist James Fisher (1912–1970) on *Fighting Man's Competitors.*[71] To varying degrees all of these episodes placed humanity at the centre of a manifest destiny, at a point whereby our species could go on to control its own evolution; an idea listeners were first introduced to by Huxley in 1925. Most broadcasts during the war that focused on, or touched upon man and evolution, also pondered upon these same questions about where human evolution will take us next.[72] And again, while generally well received, shows from this

[69] Esposito, 'Utopianism in the British Evolutionary Synthesis'.

[70] For more on the broader intellectual discussions going on during the period and their interactions with broadcasting see Timothy Boon, *Films of Fact: A History of Science in Documentary Films and Television* (Wallflower Press, 2008), chapter 4.

[71] For more on H. G. Wells and his role in broadcasting on evolutionary themes, see Chap. 5. The interested reader can listen to H. G. Wells opening episode of this series, first broadcast on 15 January 1943 online at the BBC Archive website: http://www.bbc.co.uk/archive/hg_wells/12408.shtml *(last accessed 05/06/2021).*

[72] See, for example, *Biology in the Service of Man* (1939) and *Man's Place in Nature* (1942) covered in Chap. 4. Not to be confused with *Biology in the Service of Man* (1932), from 1937 onwards this title was used for a distinct series in the "For the Schools" slot, which although hugely varied in its content, nearly always framed biological science in a utilitarian and functionalist way; for example see: For the Schools, *Biology in the Service of Man*, "Man's Origin". Written and presented by Professor Harold Munro Fox. BBC National Programme, 21 June, 1939. *Man's Place in Nature.* BBC Home Service, October–December 1942.

period that featured this progressive and at points trans-humanist narrative received complaints about their dogmatic vision of science.[73]

While varying in some aspects, all of these series borrow heavily from the narrative of the *evolutionary epic* (as introduced in Chap. 1), commonly utilised by Julian Huxley's grandfather, Thomas Henry Huxley and other nineteenth-century science popularisers. Yet despite the use of this familiar narrative arc, for many Christian audience members in the austere wartime setting of Britain, the connection of evolutionary science with aspects of scientific materialism, progressivism and humanism would have been problematic. In *Reshaping Man's Heritage*, Huxley's conjoining of biological evolution with progressive social and political ideas reinforced the idea that there was a necessary connection between evolutionary theory, secularism and other progressive, or even radical, political outlooks. For many older listeners, who may not have shared Huxley's enthusiasm for the redemptive and inevitable progressive march of science, this association between evolutionary science, secularism and radical politics would already have been familiar from late nineteenth-century debates.[74]

By the end of the war it was already becoming apparent that more than just the progressive frames being used to discuss science were alienating potential listeners, as science talks that relied on experts dryly lecturing over the airwaves were routinely failing to reach the diverse audience BBC producers desired. A review of *Reshaping Man's Heritage* in *Nature* noted this exact problem, stating that although the reviewer found the "distinguished" lecturers, the subjects and the language used all exemplary, those attending an army listening group that the reviewer had organised were not stimulated at all. The reviewer argued:

> [W]hen a straight talk is given, much more attention is paid to the speaker's degree of 'mike-worthiness' than is the case with the Adults Talks Department of the B.B.C. That a man or woman is an authority on a subject is not enough to make him a good broadcaster. He should have a microphone manner which will appeal to the largest body of listeners. … It is

[73] See, for example, Barnes, "Letter to A.V. Hill," 27 January 1943, and other letters in R51/523, BBC-WA. For complaints in response to *Man's Place in Nature*, from both scientific and religious perspectives, see R41/51/1, BBC-WA. For an overview of the series see: A. V. Hill, "In Terms of the Universe", *The Radio Times*, September 25, 1942, 4.

[74] See, for example, James R. Moore, *The Post-Darwinian Controversies: A Study of the Protestant Struggle to Come to Terms with Darwin in Great Britain and America, 1870–1900* (Cambridge University Press, 1979).

> remarkable that the number of first-rate broadcasters on any special subject number less than half a dozen. How then, can science topics be presented to make the appeal which they inherently possess for the bulk of the population? One of the answers lies in the greater use of feature programmes.[75]

Thus, after a decline in popularity in the late 1930s, attributed to Mary Adams' switch to television in 1936 and the emergence of a more conservative climate at the BBC in general, in the years following the war features broadcasts on science, and in particular science and society, were to come back into vogue.[76]

Post-war Consolidations and Contestations

In the immediate post-war years, as media restrictions were lifted and the organisation began to grow again, there was an influx of young writers and producers at the BBC. Of this fresh talent, a handful stand out in the records for their personal commitment to the science and society approach. The first of whom, Nesta Pain (1905–1995), was a recently divorced struggling playwright, who had joined the BBC in 1942 as a writer and features producer.[77] Although Pain did not come from a science background herself, early successes with shows on the tsetse fly and sleeping sickness fuelled a burgeoning interest in programmes about the application of science, particularly medicine.[78] In writing and producing science features, Pain brought many tools from her background as a playwright, emphasising the importance of allying scientific fact and accuracy with narrative structure and emotion. In a departure from nearly all prior science broadcasts at the BBC, Pain stated that when choosing the initial

[75] T.H. Hawkins, "Science and Broadcasting", *Nature*, 8 July 1944, 3897:154, 38–39.

[76] Jones, 'Mary Adams and the Producer's Role in Early BBC Science Broadcasts', 977–979.

[77] Kate Terkanian and Hugh Chignell, 'Nesta Pain', *Media History* 26:1 (2020): 20–33, https://doi.org/10.1080/13688804.2019.1679619; and Richard Hewlett, "Pain, (Florence) Nesta Kathleen (1905–1995)", Oxford Dictionary of National Biography, Oxford University Press, 2004, http://www.oxforddnb.com/view/article/58024.

[78] *War Against Disease*. Written and produced by Nesta Pain, BBC Home Service, 26 February 1944; and Forces Educational Broadcast, *War Against the Tsetse Fly*. Written and produced by Nesta Pain, BBC Light Programme, 21 September 1945.

subject for a science feature "the first criterion must be, 'is it of dramatic value?'"[79]

In 1945 Pain moved to the newly formed Forces Education Unit (FEU), and was joined by Archie Clow (1909–1996), a historian of science, science populariser and journalist from Aberdeen.[80] Clow arrived at the BBC with a vision for the role science broadcasting could play in aiding societal advance, informed by the principles of scientific humanism, as taught to him at the University of Aberdeen by his mentor and colleague of Julian Huxley, Professor Lancelot Hogben (1895–1975).[81] Clow went on to become head of science programming for BBC Radio, and throughout his 24 years at the organisation continued to be involved in the strategic management, commissioning, writing and production of a wide range of radio science features.[82]

When Clow and Pain began working together on their initial features for the FEU, likeminded colleagues interested in developing the format and scope of science broadcasting at the BBC were few and far between, scattered across departments. In particular, Clow and Pain would have found allies in educational broadcasts, where both the Director of Schools Broadcasting, Mary Somerville (1897–1963), and Richard Palmer, a Senior Assistant to Somerville, had already been reflecting on the potential

[79] Nesta Pain, "B – Science in Feature Programmes", BBC note on *Science and Radio*, a conference held at Cambridge, 18–19 May 1946, R51/529/4 – Talks. Science and Broadcasting, 1943–1946, BBC-WA. All BBC copyright content reproduced courtesy of the British Broadcasting Corporation. All rights reserved.

[80] Formed in February 1945, the Forces Educational Unit (FEU) was tasked with assisting the reintegration of soldiers returning in their thousands into civilian life. In meeting this remit many of the broadcasts produced were innovative, and the output of the FEU attempted to balance education and the development of relevant applied skills, with the motivational and morale raising role many Forces Education broadcasts had played during the war. Asa Briggs, *The History of Broadcasting in the United Kingdom: Volume IV: Sound and Vision* (OUP, 1995), 736–743. For more on FEU's role in science broadcasts on evolution, see Chap. 4.

[81] Hogben was Regius Professor of Natural History at the University of Aberdeen where Clow had completed his PhD and was a peer and colleague of Julian Huxley's; both men were founding Editorial Board members of the *Journal of Experimental Biology* in 1923, and both were on the Associated Realist Film Producers advisory committee, along with fellow biologist J. B. S. Haldane. For more, see Lancelot Thomas Hogben, *Lancelot Hogben, Scientific Humanist: An Unauthorised Autobiography* (Merlin Press, 1998).

[82] For a detailed account of Clow's career that situates his work within the wider trajectory of science broadcasting at the BBC, see Jared Robert Keller, 'A Scientific Impresario: Archie Clow, Science Communication and BBC Radio, 1945–1970', PhD Thesis, Imperial College London, 2017, http://spiral.imperial.ac.uk/handle/10044/1/57504.

roles for science broadcasting. Since joining the BBC in 1925, Somerville had been committed to experimenting with different formats for educational broadcasts.[83] Yet, while the Schools Broadcasting team were committed to situating science within its social and historical contexts, and were pioneers of using new approaches, such as dramatisation, their remit was purely educational. Thus, they diverged from Pain's criterion of using dramatic value to inform the selection of what scientific subjects should feature in broadcasts.[84]

In parallel to the early endeavours of developing formats for radio science features, those working for BBC Television, which was relaunched in the summer of 1946 after a wartime hiatus, were also grappling with the question of how to make science appealing to the masses, via this new visual medium.[85] Of particular note in this area was Aubrey Singer (1927–2007), who joined Outside Broadcasts in 1949, and by the mid-1950s was specialising in producing television science features. Singer, who had been trained in film editing at the British Gaumont film company, was part of the first cohort of staff at the BBC trained in film techniques, as opposed to radio. He went on to have a long and influential career at the BBC, becoming Head of Science and Features for television in 1961, and eventually rising to Deputy-Director General before his retirement in 1984.[86]

In the immediate post-war years, in addition to the continuing development of radio science broadcasts, and initial forays into televised science broadcasting, the mechanisms for co-ordinating science broadcasts at the BBC also began to be formalised. A debate about whether external advisors should be allowed to formally input into BBC science programming had been brewing for several years. Various scientific societies and science advocacy groups, including Huxley via the British Association for the

[83] For more on Somerville's long and pioneering career at the BBC, see Murphy, *Behind the Wireless*, 156–159.

[84] For more on Schools Broadcasting's role in the development of content on evolution see Chap. 4. Allan Jones, 'Speaking of Science: BBC Science Broadcasting and Its Critics, 1923–64' (University College London, 2010), 143–146, http://discovery.ucl.ac.uk/19988/; Keller, 'A Scientific Impresario', 65–67.

[85] Mary Adams, now a producer in the television department, brought her experience of developing science features for radio to these discussions. See Timothy Boon, *Films of Fact: A History of Science in Documentary Films and Television* (Wallflower Press, 2008), 192–196.

[86] Michael Leapman, "Singer, Aubrey Edward (1927–2007)", Oxford Dictionary of National Biography, Oxford University Press, 2011, https://doi.org/10.1093/ref:odnb/98830.

Advancement of Science (BAAS), had been petitioning the BBC to create an external Scientific Advisory Committee.[87] Furthermore, Huxley had also been trying to influence science programming from another direction, having sat on the Talks Advisory Committee since 1937.[88] While executives at the BBC valued the input of senior scientists on specific content, they strongly defended the autonomy and expertise of their own broadcasters. However, despite repeated resistance, in 1949 a discussion and review into the broadcasting of science began.[89]

By the late 1940s the science and society approach had proliferated so much that when the BBC began soliciting scientists as part of this review, the physicist Mark Oliphant, who had himself begun the discussion as a newly appointed member of the BBC's General Advisory Council, lamented:

> What I would like to see is some break-away from the perpetual theme of 'science and society', with the inevitable excursions of the scientist into fields of politics where he does not shine, towards an attempt to present science as natural philosophy, as a way of life and culture in its own right. I believe it can be done.[90]

However, the science and society approach had not only proven itself successful in reaching new audiences, but it was becoming increasingly evident that it worked well with the visual nature of television, now being rolled out nationwide. Thus, despite Oliphant and others' perhaps well-placed misgivings about the BBC's focus on applied science at the expense of "the fundamental science on which it rests" and a subsequent drive to redress this imbalance, the science and society approach continued to proliferate in the years that followed.[91] As we shall see in subsequent chapters,

[87] Jones, 'Speaking of Science', 143, 163–170.

[88] Jones, 'Mary Adams and the Producer's Role in Early BBC Science Broadcasts', 16.

[89] This review of science broadcasting was triggered by the reformulation of the BBC's General Advisory Council in 1947 to include more diverse representative interests—including more scientists—and occurred when the whole of the BBC was itself under review as part of the Beveridge Committee convened in 1949. See Boon, *Films of Fact*; Briggs, *The History of Broadcasting in the United Kingdom*, 1995, 184–191.

[90] Mark Oliphant, "The Broadcasting of Science," Memorandum to the BBC General Advisory Council, 6 May 1949, R6/34 GAC – Science Sub-Committee, BBC-WA. All BBC copyright content reproduced courtesy of the British Broadcasting Corporation. All rights reserved.

[91] Oliphant, "The Broadcasting of Science." Other figures supporting Oliphant's position included the physicist Sir Lawrence Bragg and the civil servant and politician Sir John Anderson, see: "The Presentation of Science," Report of discussion of the BBC General Advisory Council,

this tension and contestation between presenting science as value-free, neutral and transcending culture versus depicting it as just one cultural aspect of contemporary society regularly re-emerged over the subsequent decades. This interface—education versus application, fundamental science shorn of its ethical and societal implications versus depicting science in an applied (often value-laden) role within society—was often treated by scientists, and some broadcasters, as a dichotomous choice. Here we see why Huxley's utilisation of scientific humanism, with its ability to bypass such demonstrably artificial dichotomies, lent itself so well to broadcast media. By centring scientific content around a core progressive narrative, especially via the adoption of the evolutionary epic in broadcasts on the biological sciences, Huxley's vision of scientific humanism was able to side-step this tension between these two different visions for science broadcasting. Huxley's connections with elites of the scientific world kept him at the vanguard of new scientific developments and discoveries, while his eclectic intellectual interests enabled him to pitch broadcasts that situated this cutting-edge science within a broader societal context. As a more comprehensive, planned approach to science broadcasting emerged at the BBC, especially through the expansion of Clow's and Singer's work on science features for radio and television respectively, scientifically progressive narratives, such as the evolutionary epic, began to be embedded in the very fabric of the broadcast formats that were being developed.

Conclusion

In this chapter I have given an overview of the earliest radio broadcasts in the UK on evolution. Situating them within the wider development of science broadcasting in the period, I have shown how Julian Huxley influenced both the scientific content and more importantly the philosophical framing used in these early forays into radio science. I have shown how the very first broadcast on evolution, Huxley's *The Stream of Life*, was in keeping with his wider popular writings from the period and used the narrative of the evolutionary epic. In highlighting the successful transition of this science communication narrative, first utilised by nineteenth-century science popularisers, we have begun to explore the foundations of the dominant framing of evolution that emerged in later decades as the BBC developed large-scale documentaries on the subject.

2 June 1949, R6/34 GAC – Science Sub-Committee, BBC-WA. All BBC copyright content reproduced courtesy of the British Broadcasting Corporation. All rights reserved.

Following the work of Robert Bud, we have seen how the censorship of broadcast content elsewhere at the BBC, particularly in the Religion Department, led to the emergence of humanist worldviews in science broadcasts, particularly those produced by Mary Adams in the 1930s. To imprint this scientific humanist vision onto broadcasts about biological evolution, Julian Huxley turned to the narrative arc of the evolutionary epic. Thus, while he taught listeners that evolution was random and directionless, he did so somewhat paradoxically as part of a narrative structure that frequently centred on the progressive march of science, and regularly referenced the potential for humanity to manage its own future biological evolution.

While we have seen that the content of Huxley's and other science broadcasts in this mould was on occasion deemed to have overreached the desired boundaries of science content, ultimately this materialistic approach persisted and remained popular. Perhaps this is because although it was too non-spiritual and mechanistic for some, for the majority of the audience, Huxley's way of speaking about evolution felt familiar. For the narrative used is often teleological, progressive and a continuation of a natural theological tradition, both in its use of language and in its reservation of a special place for humans. It is a humanistic version of the original evolutionary epic presented in the Bible, which was perhaps more palatable than the "red in tooth and claw" version of evolution that had caused so much fear in the decades following Darwin's original publication.[92]

Thus the division did not fall where readers from the twenty-first century might presume—between religious and scientific ways of understanding the natural world—but rather that broadcasts by both religious and non-religious scientists were making an argument about scientific progressivism and the distinctness of man, that was rooted in a natural theological tradition. As we shall see in Chap. 3, the centrality of progress and the space for the distinctiveness of man in Huxley's broadcasts were something capitalised upon by accommodationist Religion Department broadcasts on evolution and religion in the 1950s.

Alongside Huxley's narrative framing of the subject, I have introduced some of the early producers who were thinking about how best to impart

[92] Attributed to the poet Alfred Tennyson the phrase "red in tooth and claw" became inextricably linked with popular British debates about evolution following Darwin's publication of *On the Origin of Species* in 1859. See Michael Ruse, *The Darwinian Revolution: Science Red in Tooth and Claw* (University of Chicago Press, 1999).

scientific knowledge via radio broadcasts in the 1920s and 1930s. In particular, Adams' creative vision—her broadening of the scope of coverage from the purely educational and didactic, towards popular features about science in society—was a crucial step in the emergence of a distinct British way of making science broadcasts. When tensions were to re-emerge in the post-war years about whether science broadcasts should stick to the popular science and society approach, or to re-orientate along more knowledge imparting and pedagogically focused lines, Huxley's approach was able to neatly bypass this argument. His broadcasts simultaneously were able to give concrete applied everyday examples, situated in a wider unifying vision and packaged within a familiar narrative, yet also supposedly remaining politically neutral because they were rooted in objective biological science. As we shall see in the following chapter, both Huxley's narrative approach and the creative format considerations as typified by Adams' early endeavours became even more important as the airtime for science broadcasting proliferated and the visual medium of television rose to prominence.

References

Adams, Sally. "Adams, Mary Grace Agnes (1898–1984)," *Oxford Dictionary of National Biography*, Oxford University Press, 2004, http://www.oxforddnb.com/view/article/30750.

BBC, *The British Broadcasting Corporation Handbook*. HMSO, 1928.

Boon, Timothy. *Films of Fact: A History of Science in Documentary Films and Television*. Wallflower Press, 2008.

———. '"The Televising of Science Is a Process of Television": Establishing Horizon, 1962–1967'. *The British Journal for the History of Science* 48:1 (2015): 87–121. https://doi.org/10.1017/S0007087414000405.

Bowler, Peter J. *Science for All: The Popularization of Science in Early Twentieth-Century Britain*. University of Chicago Press, 2009.

Briggs, Asa. *The History of Broadcasting in the United Kingdom: Volume I: The Birth of Broadcasting*. Oxford University Press, 1961.

———. *The History of Broadcasting in the United Kingdom: Volume II: The Golden Age of Wireless*. Oxford University Press, 1965.

———. *The History of Broadcasting in the United Kingdom: Volume IV: Sound and Vision*. Oxford University Press, 1995.

Bud, Robert. '"The Spark Gap Is Mightier than the Pen": The Promotion of an Ideology of Science in the Early 1930s'. *Journal of Political Ideologies* 22:2 (2017): 169–181. https://doi.org/10.1080/13569317.2017.1298548.

Butler, Samuel. "Darwin among the Machines," *The Press*, Christchurch, New Zealand, June 13, 1863.
Cain, Joe. 'Julian Huxley, General Biology and the London Zoo, 1935–42'. *Notes and Records of the Royal Society* 64:4 (2010): 359–378. https://doi.org/10.1098/rsnr.2010.0067.
Cole, Rupert. '1972: The BBC's Controversy and the Politics of Audience Participation'. *Public Understanding of Science* 26:4 (2017): 514–518. https://doi.org/10.1177/0963662516684231.
Church, Archibold. "Editorial", *The Realist*, 1:1, April 1929, 180.
Crompton, Richmal. *William and The Brains Trust.* Macmillan Children's Books, 1989.
Dunn, Robert G. 'Science, Technology and Bureaucratic Domination: Television and the Ideology of Scientism'. *Media, Culture & Society* 1:4 (1979): 343–354. https://doi.org/10.1177/016344377900100403.
Esposito, Maurizio. 'Utopianism in the British Evolutionary Synthesis'. *Studies in History and Philosophy of Science Part C: Studies in History and Philosophy of Biological and Biomedical Sciences*, Defining Darwinism: One Hundred and Fifty Years of Debate, 42:1 (2011): 40–49. https://doi.org/10.1016/j.shpsc.2010.11.007.
Fahy, Declan, 'The Celebrity Scientists: A Collective Case Study', PhD Thesis (Dublin City University, 2010).
Falby, Alison. "Heard, Henry Fitzgerald [Gerald] (1889–1971)," *Oxford Dictionary of National Biography*, Oxford University Press, 2013, https://doi.org/10.1093/ref:odnb/54040.
Gouyon, Jean-Baptiste. 'From Engaged Citizen to Lone Hero: Nobel Prize Laureates on British Television, 1962–2004'. *Public Understanding of Science* 27:4 (2018): 446–457. https://doi.org/10.1177/0963662518760790.
Hawkins, T. H. "Science and Broadcasting", *Nature*, 8 July 1944, 3897:154, 38–39.
Heard, Gerald. *The Ascent of Humanity: An Essay on the Evolution of Civilization from Group Consciousness Through Individuality to Super-Consciousness.* Harcourt Brace, 1929.
Heard, Gerald. *Science in the Making.* Faber and Faber, 1935.
Hewlett, Richard. "Pain, (Florence) Nesta Kathleen (1905–1995)", Oxford Dictionary of National Biography, Oxford University Press, 2004, http://www.oxforddnb.com/view/article/58024.
Hogben, Lancelot Thomas. *Lancelot Hogben, Scientific Humanist: An Unauthorised Autobiography.* Merlin Press, 1998.
Huxley, J. S. *Reshaping Man's Heritage; Biology in the Service of Man: By J. S. Huxley, H. G. Wells, J. B. S. Haldane … [et al.].* Allen & Unwin, 1944.
Huxley, Julian. *Essays of a Biologist.* A. A. Knopf, 1923. http://archive.org/details/essaysofbiologis1923huxl.
———. *Evolution, the Modern Synthesis.* Allen & Unwin, 1942.

———. *The Stream of Life*. Watts & Co., 1926.
Jones, A. C. 'Speaking of Science: BBC Science Broadcasting and Its Critics, 1923–64'. PhD Thesis (University College London, 2010). http://discovery.ucl.ac.uk/19988/.
Jones, Allan. 'Elite Science and the BBC: A 1950s Contest of Ownership'. *The British Journal for the History of Science* 47:4 (2014): 701–723. https://doi.org/10.1017/S0007087413000927.
———. 'Mary Adams and the Producer's Role in Early BBC Science Broadcasts'. *Public Understanding of Science* 21:8 (2012): 968–983. https://doi.org/10.1177/0963662511419450.
Julian Huxley. *The Uniqueness of Man*. Chatto & Windus, 1943. http://archive.org/details/TheUniquenessOfMan.
Kamminga, Harmke, and Geert Somsen. *Pursuing the Unity of Science: Ideology and Scientific Practice from the Great War to the Cold War*. Routledge, 2016.
Keller, Jared Robert. 'A Scientific Impresario: Archie Clow, Science Communication and BBC Radio, 1945–1970', PhD Thesis (Imperial College London, 2017). http://spiral.imperial.ac.uk/handle/10044/1/57504.
Knight, David M. and Matthew D. Eddy, *Science and Beliefs: From Natural Philosophy to Natural Science, 1700–1900*. Routledge, 2017.
Leapman, Michael. "Singer, Aubrey Edward (1927–2007)", Oxford Dictionary of National Biography, Oxford University Press, 2011, https://doi.org/10.1093/ref:odnb/98830.
Matheson, H. *Broadcasting*. Butterworth, 1933.
Moore, James R. *The Post-Darwinian Controversies: A Study of the Protestant Struggle to Come to Terms with Darwin in Great Britain and America, 1870–1900*. Cambridge University Press, 1979.
'Mr. Eldon Moore.' *The Times*, November 22, 1954.
Murphy, Kate. *Behind the Wireless: A History of Early Women at the BBC*. Springer, 2016.
Olby, Robert. "Huxley, Sir Julian Sorell," *Oxford Dictionary of National Biography*, 2004, https://doi.org/10.1093/ref:odnb/31271.
Parrish, J. M. and John R. Crossland (Eds), *The Science of Living Things*, The Standard University. Odhams Press Ltd, 1935.
Reiser, Oliver. *The promise of scientific humanism toward a unification of scientific, religious, social and economic thought*. University of Wisconsin, 1940.
Reith, John. *Broadcast over Britain*. Hodder and Stoughton, 1924.
Robert Westman. 'The "Two-Cultures" Question and the Historiography of Science in the Early Decades of the Salk Institute for Biological Studies'. *Sartoniana* 32 (2019). http://www.sartonchair.ugent.be/file/293.
Ruse, Michael. *The Darwinian Revolution: Science Red in Tooth and Claw*. University of Chicago Press, 1999.
Singer, Charles. "Scientific Humanism", *The Realist*, 1:1, April 1929, 12–18.

Smocovitis, Vassiliki Betty. *Unifying Biology: The Evolutionary Synthesis and Evolutionary Biology*. Princeton University Press, 1996.
“Some Coming Wireless Talks: The King’s Speech,” *The Manchester Guardian*, September 26, 1925, 13
Street, Sean. *Historical Dictionary of British Radio*. Rowman & Littlefield, 2006.
Terkanian, Kate, and Hugh Chignell. ‘Nesta Pain’. *Media History* 26:1 (2020): 20–33. https://doi.org/10.1080/13688804.2019.1679619.
Waters, C. Kenneth. *Julian Huxley, Biologist and Statesman of Science: Proceedings of a Conference Held at Rice University, 25–27 September 1987*. Rice University Press, 1992.
Weindling, Paul. ‘“Julian Huxley and the Continuity of Eugenics in Twentieth-Century Britain”’. *Journal of Modern European History* 10:4 (2012): 480–499. https://doi.org/10.17104/1611-8944_2012_4.

CHAPTER 3

Evolving Content for the Small Screen, from Radio to Early Television Formats

In 1949, during the review of broadcasts on science introduced in the previous chapter, the BBC received an unsolicited application from one of their more sophomoric science broadcasters, the mathematician Jacob Bronowski. He offered his services in a new, self-proposed role as a BBC science adviser. Bronowski had first appeared on BBC radio in 1946, and in his letter he stated: "[i]f non-scientists are to blame for their neglect of science, then scientists cannot escape as great a blame for their ignorance of literature, the arts, and the general place of their work in culture and in history".[1] Unsurprisingly, given the difference between Bronowski's view and that being put forward by the initiator of the review process, the physicist Mark Oliphant—who wished to "present science as natural philosophy"[2]—Bronowski was politely turned down by the BBC.[3]

[1] Correspondence from Jacob Bronowski to George Barnes, 29 May 1949, R6/34, BBC-WA. All BBC copyright content reproduced courtesy of the British Broadcasting Corporation. All rights reserved.

[2] See Chap. 2; and Mark Oliphant, "The Broadcasting of Science," Memorandum to the BBC General Advisory Council, 6 May 1949, R6/34 GAC – Science Sub-Committee, BBC-WA. All BBC copyright content reproduced courtesy of the British Broadcasting Corporation. All rights reserved.

[3] Correspondence from George Barnes to Jacob Bronowski, 25 January 1950, R6/34, BBC-WA. For more on wider discussion of appointing a scientific advisor in this period see Timothy Boon, *Films of Fact: A History of Science in Documentary Films and Television* (Wallflower Press, 2008), 189–191.

A. Hall, *Evolution on British Television and Radio*, Palgrave Studies in Science and Popular Culture,
https://doi.org/10.1007/978-3-030-83043-4_3

Against the backdrop of the BBC's review of science content, this chapter charts the ongoing deliberations between the communications experts of the BBC, their allies such as Bronowski and external institutional and more conservative influences trying to shape science broadcasts in the post-war years. Exploring how this tension effected the construction of shows on evolution, the chapter outlines a broadening of approaches and content, beginning with the first detailed "accommodationist" account on the relationship between evolutionary theory and Christian theology, *Science and the Christian Man* (1951). The chapter then outlines how, in response to complaints from anti-evolutionist religious groups, the science producer Archibald Clow drew up the institutions' first, and only, editorial guidelines for content on evolution. The chapter finishes with the first television broadcast on evolution, 1958's *Five Hundred Million Years*, reflecting on the continuities and differences between earlier radio broadcasts on the subject, and the plethora of emerging ways of presenting science on the new visual medium.[4] First, we return to Jacob Bronowski and his overlooked role in expanding the scope of science broadcasting in Britain.

Expanding the Cast

A mathematician by training, the Polish-born intellectual Jacob Bronowski (1908–1974) had a wide range of scholarly interests and held a diverse range of posts across his remarkable and varied career. Immigrating with his family to England at the age of 12, he received his doctorate from the University of Cambridge in 1933, and began lecturing at the University of Hull in 1934. During the war, Bronowski was seconded to the Ministry of Home Security's Research and Experiments Department, where he carried out statistical research to improve the performance of bombing raids on German industrial targets.[5] In November 1945, Bronowski co-led an expedition of British civilian scientists and engineers to examine the atomic bomb sites in Hiroshima and Nagasaki, subsequently publishing a report

[4] This show also features in Chap. 7 on the commemoration of Darwin and evolution in 1958/9 and 2009.

[5] Ralph Desmarais, 'Jacob Bronowski: A Humanist Intellectual for an Atomic Age, 1946–1956', *The British Journal for the History of Science* 45:4 (2012): 576, https://doi.org/10.1017/S0007087412001069. For a popular biographical account of Bronowski's life and career see: Timothy Sandefur, *The Ascent of Jacob Bronowski: The Life and Ideas of a Popular Science Icon* (Prometheus Books, 2019).

on the atrocities for the British Government.[6] The experience would go on to shape his approach as a public intellectual, championing the positive role that science can play in society (Chap. 6). Indeed, it was Bronowski's role in the British mission to Japan that led to his very first broadcast with the BBC. On June 30, 1946, to coincide with the Home Office release of his report on the mission, he gave a live talk about his experiences in Hiroshima and Nagasaki.[7] His conversational style was an immediate success. In addition to continued appearances speaking about the issues raised by the bomb and atomic energy more generally,[8] Bronowski quickly became a go-to expert for a wide range of subjects, broadcasting a talk about the poet William Blake and joining *The Brains Trust* panel for the first time in October 1946.[9] Asa Briggs, institutional historian of the BBC, would later describe Bronowski as, "one of the most accomplished and versatile broadcasters ever employed by the BBC".[10]

As well as having a firm commitment to bridging the divide between the arts and the sciences, Bronowski's earliest broadcasts were infused with the ideology of scientific humanism, as first iterated on the BBC by Julian Huxley and his humanist peers in the 1930s.[11] Bronowski was a

[6] Jacob Bronowski, *The Effects of the Atomic Bombs at Hiroshima and Nagasaki: Report of the British Mission to Japan* (His Majesty's Stationery Office, 1946).

[7] Desmarais, 'Jacob Bronowski', 577–578.

[8] Notable in this regard were Bronowski's radio plays *Journey to Japan* (1948) and *The Face of Violence* (1950); the latter sharing the 1950 Prix Italia, an esteemed and long-running award for outstanding broadcasts awarded annually by the Italian national broadcaster, Radiotelevisione Italiana. Asa Briggs, *The History of Broadcasting in the United Kingdom: Volume IV: Sound and Vision* (OUP, 1995), 507; Ralph John Desmarais, 'Science, Scientific Intellectuals, and British Culture in the Early Atomic Age: A Case Study of George Orwell, Jacob Bronowski, P.M.S. Blackett and J.G. Crowther' (Imperial College London, PhD Thesis, 2010), http://spiral.imperial.ac.uk/handle/10044/1/5646, 78.

[9] The talk on William Blake was the first in a series of three written and presented by Bronowski, in the other two talks in the *Invention and Imagination* series Bronowski spoke about Humphry Davy and Thomas Jefferson (*Invention and Imagination*, BBC Third Programme, Oct to Dec 1946). Bronowski was a keen scholar of William Blake, and held a long-standing interest in poetry, in his first major publication, *The Poet's Defence* (Cambridge University Press, 1939) Bronowski had expounded the ability of poetry to tell truths unascertainable by science. Desmarais, 'Jacob Bronowski', 575–576.

[10] Briggs, *The History of Broadcasting in the United Kingdom*, 507.

[11] The Beveridge Broadcasting Committee of 1949 had recommended organisations that studied moral or spiritual issues, but were not religious should have the opportunity to broadcast, but only via the Talks Department rather than the Religious Broadcasting Department. In 1958 The Humanist Association complained that this recommendation was

younger contemporary of Huxley, and both increased their popularity among the British public via their appearances on *The Brains Trust*.[12] The career path pioneered by Huxley provided a route for Bronowski to follow, and both figures are central to the popularisation of scientific humanism in Britain in the first half of the twentieth century.[13] In a radio talk on "Unbelief and Science" aired in 1948 as part of a series on the *Ideas and Beliefs of the Victorians*,[14] Bronowski argued that, "[Thomas Henry] Huxley and Clifford and Tyndall and their fellow scientists had undermined the popular basis of belief almost unnoticed and left it hollow and ready for a landslide".[15] In this and other radio talks he gave during the period, most notably his series on *The Common Sense of Science* (1948), Bronowski used the history of science, and specifically the reception of Darwinian evolutionary theory, to discuss controversy, introducing a clash narrative between religious institutions and science. Although *The Common Sense of Science* was ostensibly about the history of the cultural divide between science and art, Bronowski also explained how the "quarrel between science and soul was trumped up by the religious apologists … who were anxious to find science materialistic and unspiritual".[16] Although by today's standards the historiographies presented in these post-war

not being implemented. Penry Jones, "The BBC and the Humanists," BBC Paper presented to CRAC, 24 February 1969, R6/21/10, BBC-WA.

[12] Bronowski first appeared on *The Brains Trust* panel on October 8, 1946, as a stand in for philosopher and regular panellist Cyril Joad. While Bronowski continued with occasional appearances on the radio show until it ended in 1949, it was when it relaunched on television in 1955 that he became a regular, making over 80 appearances including the final episode, a transatlantic special, broadcast on 2 November 1961. On occasion Huxley and Bronowski did appear alongside each other on *The Brains Trust* panel, for an example see *The Brains Trust*, 11 September 1955, BBC TV. For more on the *Brains Trust* see Chap. 2.; *Radio Times*, October 4, 1946, 10; and Desmarais, 'Science, Scientific Intellectuals, and British Culture in the Early Atomic Age', 36 and 72.

[13] Robert Bud, '"The Spark Gap Is Mightier than the Pen": The Promotion of an Ideology of Science in the Early 1930s', *Journal of Political Ideologies*, 22:2 (2017), https://doi.org/10.1080/13569317.2017.1298548; Desmarais, 'Jacob Bronowski'.

[14] Julian Huxley also gave a talk in this series: *Ideas and Beliefs of the Victorians*. "Evolution and Human Progress." Produced by Anna Kallin. Written by Julian Huxley. BBC Third Programme, March 22, 1948.

[15] Interestingly the listing in the *Radio Times* for the repeat of the same show later that evening gave it the title: "Religion and Evolution". *Ideas and Beliefs of the Victorians*. "Unbelief and Science." Produced by Anna Kallin. Written by Jacob Bronowski. BBC Third Programme, 17 March 1948, script on microfilm at BBC-WA. All BBC copyright content reproduced courtesy of the British Broadcasting Corporation. All rights reserved.

[16] Jacob Bronowski, *The Common Sense of Science* (Faber & Faber, 2011), 9.

broadcasts were "Whiggish",[17] they were the BBC's first broadcasts on evolution that went beyond the science and scientists to discuss the theories' relationship to popular culture and other cultural institutions, such as religion.[18]

Unlike Bronowski's expansive historical sketches that connected enlightenment thinkers with Victorian men of science, the few pre-war instances that covered the history of evolutionary theory were limited to individual biographical accounts.[19] Bronowski's more in-depth coverage of the societal impacts of nineteenth-century evolutionary science was afforded by the wide-ranging scope of *Ideas and Beliefs of the Victorians*, which broadcast 52 episodes over a 4-month period. The series dedicated a significant proportion of airtime to considering religion and man's place in the world, in specific response to evolution. The episode prior to Bronowski's, "Grounds for Unbelief", discussed George Eliot's agnosticism, while the episodes following him saw Regius Professor of Divinity at Cambridge, Canon Charles E. Raven, speak on "Man and Nature", and Julian Huxley speak about "Evolution and Human Progress".[20] Although promoted by the BBC as presenting a well-rounded collection of perspectives, including those on opposing sides of nineteenth-century arguments, none of the intellectuals who presented the series argued for an a priori rejection of evolution.

[17] For more on Whiggish narratives in the history of science see: David Alvargonzález, 'Is the History of Science Essentially Whiggish?', *History of Science* 51:1 (2013): 85–99, https://doi.org/10.1177/007327531305100104.

[18] There were at least two pre-war broadcasts that discussed evolution and religion; however, both took an accommodationist approach, and were only broadcast regionally. The first was an episode of a cultural anthropology series by the proponent of the British Evolutionary School, R.R. Marrett, and the second was an episode of the autobiographical talks show *Myself and Life* by the Bishop of Birmingham who was also a mathematician and Fellow of the Royal Society. *The Making of Man*, "Religion". Written and presented by Robert Ranulph Marrett. BBC London (2LO), 10 February 1927; and *Myself and Life*, "The Bishop of Birmingham". Written and Presented by the Very Rev. Dr E. W. Barnes. BBC Regional Programme, Midlands, 4 February 1939.

[19] For example see *Biology in the Service of Man*, "Huxley (1825–1895)—Knight-Errant of Darwinism". Written by J. Arthur Thomson. BBC National Programme, June 24, 1932.

[20] *Ideas and Beliefs of the Victorians*. "Grounds for Unbelief." Written and presented by Humphry House. BBC Third Programme, March 15, 1948; *Ideas and Beliefs of the Victorians*. "Man and Nature." Written and presented by Reverend Charles E. Raven. BBC Third Programme, March 21, 1948; and *Ideas and Beliefs of the Victorians*. "Evolution and Human Progress." Written and presented by Julian Huxley. BBC Third Programme, March 22, 1948.

As discussed in Chap. 2, whenever the BBC broadcast a programme about evolution, there were a small number of letters of complaint objecting on religious grounds.[21] Given that estimated viewing figures for broadcasts such as Bronowski's *The Common Sense of Science* were often in the hundreds of thousands, these written complaints were in the extreme minority, and almost always came from known anti-evolutionist groups, in particular the Evolution Protest Movement, founded in 1932.[22] Despite the Director of Religious Broadcasting's assertion that less than 2% of Christian scholars and clergy in the UK contested evolution, this trickle of complaints, and the demands for airtime for opposing views, continued into the early 1950s.[23]

It is into this contested landscape that the first broadcasts presenting an accommodationist position on the relationship between evolutionary theory and Christian theology were broadcast. In a series of six talks, *Science and the Christian Man* (1951), the Church of England clergyman and Regius Professor of Divinity at Cambridge, Canon Charles E. Raven, outlined his vision for "living out the Christian faith in an age of science".[24] In the opening of episode one, Raven dismissed as "crude materialism", the nineteenth-century "God who could only be fitted into the few gaps left". He stated that "to explain life in terms of biochemistry is to exclude all the really important experiences from it", and added that his talks would address the implications of modern science for experiences such as love and the soul, which such a mechanistic worldview couldn't explain.[25] However, in his opening gambit, he also made it clear, especially with regard to evolution, that this was no anti-science diatribe:

[21] For many examples see: "Evolution: How Things Began, File 1, 1938–1945", R41/51/1 and "Evolution: How Things Began, File 1, 1946–1950", R41/51/1, BBC-WA. The letters were often part of coordinated campaigns, and the BBC received them to coincide with letters, statements and opinion articles published by the same authors in popular magazines and newspapers. See, for example, Douglas Dewar, "Letters: Broadcasting," *The Daily Sketch*, May 22, 1944.

[22] Today known as the Creation Science Movement, for more on their history see: Ronald L. Numbers, *The Creationists: From Scientific Creationism to Intelligent Design* (Harvard University Press, 2006), Chapter 8.

[23] Rev. J. W. Welch, Letter to H. Walker, September 14, 1946, R41/51/2, BBC-WA.

[24] Charles Earle Raven, *Science and the Christian Man* (SCM Press, 1952), inside cover. In addition to the broadcasts and the 1952 book, between 19 July 19 and August 23, 1951, the BBC also published the talks verbatim in their magazine *The Listener*.

[25] Raven, 8–9.

> [W]e had better look for a moment at the question of evolution—because if you do not agree with me that creation is a continuous process and that what we call evolution is broadly speaking true, then these talks will not appeal to you. I know that some Christians argue that because Darwin's original theories have been to some small extent modified, therefore we can return to the old belief that at certain points God has intervened in a series of acts of creation. ... But these points of view seem to me demonstrably wrong.

Raven went on to use the scientific consensus on evolution to support a theological position that dismissed the arbitrary divine intervention of the Old Testament:

> In the course of evolution new levels have been reached, but these do not necessarily involve breaks in the process or divine interventions: God is in the whole, if he is in it at all: we have got rid of the idea of a master-magician and come to appreciate the New Testament concept of an ever-working, ever-loving deity.[26]

Rooting man's "novelty" in his self-consciousness, the series went on to cover aspects of Christian theology that Raven considered to have been most affected by recent scientific discoveries; notably, evolution featured in three of the six episodes.[27] Raven was part of a pre-war modernist movement within the church, which had seen him and other progressive religious figures isolated by the Church of England.[28] Yet in the years after the war, with memories of the conflict still raw, many of the British radio-listening public welcomed the progressive but tolerant message at the heart of Raven's broadcasts.[29]

[26] Raven, 9–10.

[27] Despite the conciliatory tone and the support for evolutionary science in general, it should be noted that Raven was an advocate for non-Darwinian theories and the theistic evolution of Pierre Teilhard de Chardin. Peter Bowler, "The Specter of Darwinism: The Popular Image of Darwinism in Early Twentieth Century Britain" in Abigail Lustig, Robert J. Richards, and Michael Ruse, *Darwinian Heresies* (Cambridge University Press, 2004), 61–62.

[28] Peter J. Bowler, *Reconciling Science and Religion* (University of Chicago Press), 277–286.

[29] As is typical for radio broadcasts in the period, few if any newspapers reviewed Raven's lectures. For a favourable review of the book version of Raven's lectures see: Artifex, "Science and the Christian Man", *The Manchester Guardian*, March 24, 1952.

Despite their differing views on the supernatural, in *Science and the Christian Man* we can sense much of the conciliatory tone that is also evident in Bronowski and Julian Huxley's post-war broadcasts. Indeed, Raven's broadcasts were in contrast to the majority of the more traditional and conservative post-war output of the BBC Religion Department.[30] However, in Raven's broadcasts and the few other occasions when religious broadcasting intersected with the realm of science, there was clear common ground with the progressive vision of science and humanity also being forwarded in secular science broadcasts by figures such as Huxley and Bronowski.[31] Raven described the "awareness of the universe and the joy and humility that accompany it, if in its primitive form" as "the distinctive novelty which separates man from the beasts", and "at the higher levels of human achievement the source of inspiration for the artist, the thinker and the saint".[32] As we shall see in subsequent chapters, this sentiment, with its roots in natural theology, remained central to some of the most popular broadcasts on evolution, not only through the appropriation of the language of awe and spectacle, but also through the very formats developed, which relied on progressive human-centric narratives. Raven, Huxley and Bronowski's broadcasts were part of wider media and societal trends in post-war Britain, which briefly saw spaces open up for more reflective public discussion of ideas such as national identity, post-war reconstruction and the role of science in society.[33]

[30] For an example of a more traditional religious broadcast from this period, see "Art, Religion, and Contemporary Life" a talk on a conference at the Ecumenical Institute near Geneva, written and presented by R. H. Ward, BBC Third Programme, June 16, 1950. For more on the general context of the Religion Department in the period, and in particular contested attempts in the post-war years to broaden the scope of religious broadcasts, and to feature controversy in broadcasts see, Asa Briggs, *The History of Broadcasting in the United Kingdom: Volume IV: Sound and Vision* (OUP, 1995), 696–725.

[31] In addition to Raven's talks, see also a loosely connected series of discussions on science and religion, broadcast sporadically between 1950 and 1952 on the BBC Home Service, notably Bronowski featured on one episode in December 1952. "Fifty-One Society. Science, Religion and Human Values". BBC Home Service North, December 29, 1952.

[32] Raven, *Science and the Christian Man*, 11–12.

[33] Examples in this vein from the world of broadcast media include the *Reshaping Man's Heritage* (1943) series introduced in the last chapter and the 1952 *Reith Lectures*, in which Professor of Anatomy at University College London, J. Z. Young, spoke at length about "Doubt and Certainty in Science". Despite these elite discussions happening in public forums, as Cold War pressures continued to escalate in the post-war years, the average citizen's ability to input into and influence such discussions remained minimal. For more on the general direction of organised science in this period, see Jon Agar, *Science in the 20th Century and Beyond* (Polity, 2012), 263–329.

Raven's series *Science and the Christian Man* remained an outlier during the 1950s, as nearly all radio broadcasts that dealt with evolutionary science during the decade were lectures or talks given by scientists. While Huxley continued to feature,[34] a widening cast of experts from an increasingly diverse range of scientific disciplines, including botany, genetics and archaeology joined him.[35] It is in this context of a diversifying cast of scientific experts speaking about evolution and the continued complaints by fundamentalist Christian groups that in 1952 Archibald Clow undertook research into the challenges presented by broadcasting evolutionary science.

Constructing an Official Institutional Position on Evolution

The Controller of Talks, Mary Somerville asked Clow, now a Science Producer in the BBC Talks Department, to write a report on "the state of the Evolution theory" in June 1952.[36] In the report, the only in-depth review into the broadcasting of evolutionary science ever conducted by the BBC, Clow attempted to look at evolution "in as detached a way as possible", and "not from a biological or Christian point of view". Despite this assurance, Clow, who had since the mid-1940s wanted the BBC to take a firm line with these "maniacs and cranks", was not afraid of stating where his loyalties lay.[37] In outlining the many reasons why BBC staff needed a good grasp of contemporary evolutionary theory, Clow included "attacks by anti-evolutionists <u>via minutae</u>".[38] Unsurprisingly given Clow's background in chemistry and his interests in the history of science, the

[34] See, for example, his 1951 series of six radio talks *The Process of Evolution*.

[35] See, respectively, "Evolution and the Ferns". Written and presented by Peter Bell, Lecturer in Botany at the University of London, BBC Third Programme, February 18, 1952; "The Evolving Animal". Written and presented by Conrad Hal Waddington, Professor of Animal Genetics at the University of Edinburgh, BBC Third Programme, October 26, 1952; and "Social Evolution". Written and presented by Gordon Childe, Director of the Institute of Archaeology, BBC Third Programme, April 23, 1952.

[36] Internal Memorandum from Mary Somerville (Controller, Talks) to Assistant Controller Third Programme, "Third Programme Science Talks", June 13, 1952, R51/523/7, BBC-WA. All BBC copyright content reproduced courtesy of the British Broadcasting Corporation. All rights reserved.

[37] N.S. Laker, Memo to H.P.C.S., 16 July 1945, R41/51/2, BBC-WA. All BBC copyright content reproduced courtesy of the British Broadcasting Corporation. All rights reserved.

[38] Dr Archibald Clow, "Report on Evolution," March 11, 1952, 93HD 29.2, Royal Society Archive, 2. *Underline on original.*

report was written at an advanced level, both scientifically—for example, discussing gene complexes—and epistemologically—referring to issues such as proof in science versus proof in logic, and the deductive element of scientific enquiry. Notwithstanding its sophistication, the report was a masterclass in clear science communication, written in language easily accessible to other staff in the Talks Department. After briefly outlining Darwin's major findings and proposed mechanisms, Clow spent most of the report detailing the major developments since Darwin, reassuring BBC production staff that it was normal and acceptable to state on air that Darwin got some things wrong, such as his "sweeping presumptions about sexual selection". Clow covered the latest understandings from a range of areas, such as Mendelian inheritance and genetics, but unlike Huxley's published writings in the same period, he stopped short of any attempt to synthesise or theorise across disparate areas of evolutionary biology.[39] It is also worth noting that Clow did not get everything completely correct with regard to the accepted scientific knowledge of the day. In particular, in a list of cases of indirect evidence that broadcasters might turn to for a burden of proof that evolution is credible science, he cited the by-then discredited recapitulation theory.[40]

Once he had covered the "how" of evolutionary science, Clow finished by reflecting on the philosophical question of "why". After reiterating that "virtually all biologists and a significant proportion of churchmen" considered evolution a fact, he stated that once we cross over into consideration of causation, we are entering the realm of opinion.[41] Clow then offered several theories, which attempted to address the question of why

[39] Huxley's attempt to unify the disparate areas of biological study, *Evolution: the Modern Synthesis*, in which he set out to give an overview of the connections between natural selection and evolutionary progress, was published in 1942. Based on his 1936 presidential address to the Zoology Section of the British Association for the Advancement of Science, *The Modern Synthesis* was popular enough for a second edition, with an updated introduction by Huxley added in 1955. Julian Huxley, *Evolution: the Modern Synthesis* (Allen & Unwin, 1942). For more on Huxley's book, its reception and influence see: Vassiliki Betty Smocovitis, *Unifying Biology: The Evolutionary Synthesis and Evolutionary Biology* (Princeton University Press, 1996).

[40] Recapitulation Theory proposed that during gestation, the embryo went through stages that resembled the successive evolution of the species' ancestors. By the 1950s, biological scientists had largely discredited and discarded the theory, although in popular culture it would remain popular for several decades. For more see: Nick Hopwood, *Haeckel's Embryos: Images, Evolution, and Fraud* (University of Chicago Press, 2015).

[41] Clow, "Report on Evolution," 10–12.

we evolve. After outlining special creation, divine guidance, teleology, entelechies, élan vital and Bergsonian Creation Evolution, Clow stated his own preference was for Natural Selection alone, with "no moral flavour".[42] Despite his personal position, the clear distinction Clow wished to draw BBC staff's attention to was between those who investigated these questions in good faith, and those who outright rejected the weight of scientific evidence. "I have no hesitation in saying that science has evidence to refute fundamentalist argument, but the reply to the question, why does living matter evolve, is a matter of belief, and among scientists qua human beings there are shades of belief".[43]

Although the report avoided any prescriptive guidelines for broadcasts on evolution, Clow was explicit about two things. Firstly, producers should not waste airtime pandering to those with extreme a priori positions on evolution: "[f]undamentalist opposition is based on the literal interpretation of Holy Writ and if that is an initial premise then it is futile even to present the evidence for evolution because it will contradict that which is already accepted".[44] Secondly, given all of the evidence amassed since "initial Darwinism", broadcasters should be confident in allowing a platform to "a great multitude of thinking men and women who are earnestly seeking truth and finding it in a belief in evolution as a fact coupled with belief in teleology, creative evolution, etc., according to choice". Although he accepted that conversation across these groups could result in intelligent discussion, Clow finished with the caution, "that it might prove sterile since the various groups hardly talk the same language".[45]

As with other internal reports and memoranda in this period, it is unclear how widely circulated Clow's report was; however, it is likely that most in the Radio Talks Department, particularly those who had responsibility for commissioning or producing science content, read it. Whether

[42] While Clow's position is similar to Julian Huxley's in this period, as the historian of biology and philosophy Emily Herring has demonstrated, despite his public distancing from Bergson and the concept of élan vital, Bergson's work remained an important influence on Julian Huxley throughout his career. In particular, Herring argues that Huxley's desire to explore evolution as a universal concept, and his interest in the meaning of human life were greatly influenced by Bergson's metaphor of the élan. Emily Herring, '"Great Is Darwin and Bergson His Poet": Julian Huxley's Other Evolutionary Synthesis', *Annals of Science* 75:1 (2018): 40–54, https://doi.org/10.1080/00033790.2017.1407442.

[43] Clow, "Report on Evolution," 12.

[44] Clow, "Report on Evolution," 10.

[45] Clow, "Report on Evolution," 13.

it was circulated beyond Talks to other departments or not, this report is one of several during the period that were part of a process of formalising BBC science broadcasts. We do know that the BBC's new Science Adviser, Sir Henry Dale (appointed in 1950), was sent a copy of Clow's report, and in turn forwarded it on to the evolutionary embryologist Gavin de Beer.[46] While Dale received a copy, it is not evident that Clow corresponded with Dale or any other scientists when writing the report, relying largely on his own interpretation of published literature. As introduced in the previous chapter, Clow's process for this report should be viewed in the context of wider contestations over BBC autonomy with regard to the content of science broadcasting.

The sub-committee tasked with reviewing science broadcasting in response to Mark Oliphant's concern on the balance of science output had reported back to the BBC General Advisory Council in November 1949. Echoing concerns first raised by the Talks Department in the 1930s (Chap. 2), the sub-committee dedicated a significant proportion of their report to the question of audience comprehension. Incorporating results from a listener survey conducted by the Education Department in 1949, the sub-committee found that while most listeners remained interested in the broadcast, only those with a "good background of scientific knowledge" could understand the content.[47] The sub-committee stopped short of earlier calls to create a Scientific Advisory Committee, which would operate along the lines of the Central Religious Advisory Committee (CRAC),[48] instead recommending that the BBC appoint a Science Adviser of senior scientific standing who could be consulted on an ad-hoc basis. The BBC considered and rejected Bronowski among others, including the physical

[46] Letter from Sir Henry Dale (BBC Science Advisor) to Archibald Clow (Science Producer, Talks), March 18, 1952, 93HD 29.2, Royal Society Archive. De Beer was a friend and sometime collaborator of Julian Huxley's, the two having co-authored the book *The Elements of Experimental Embryology* (Cambridge University Press) in 1934.

[47] BBC General Advisory Council Report of Special Sub-Committee to consider Broadcasts on Science, November 23, 1949. R6/34, BBC-WA. All BBC copyright content reproduced courtesy of the British Broadcasting Corporation. All rights reserved. As quoted and detailed in Boon, *Films of Fact*, 189. For a more in-depth analysis of these audience comprehension experiments see Allan Jones, 'Clogging the Machinery: The BBC's Experiment in Science Coordination, 1949–1953', *Media History* 19:4 (2013): 436–449, https://doi.org/10.1080/13688804.2013.844892.

[48] Letter from Robert H. Pickard (Chairman, The Joint Council of Professional Scientists) to the Director-General BBC, February 8, 1945; and the memo by M. E. Nicholls (Senior Controller, BBC), August 15, 1945, R6/288/1, BBC-WA.

chemist and novelist C. P. Snow,[49] before appointing Sir Henry Dale, physiologist and former President of the Royal Society (1940–1945), to the role of Science Adviser in July 1950.[50]

While Dale was knowledgeable on a broad range of scientific subjects, and commanded the ear of a wide network of scientists, he was ultimately a conservative appointment. At 75, he did not have a good grasp on the up-and-coming scientists who often made the best broadcasters, especially with regard to the new medium of television.[51] With little time to dedicate to the role, and without any BBC staff to assist him, other than the enthusiasm of an already overworked Archibald Clow, Dale was unable to make any significant inroads against his terms of reference. Although the BBC renewed his initial appointment after a year, in 1952 they quietly allowed the arrangement to lapse.

While BBC executives contemplated whether to persist with the role, Dale wrote a report on his experience. From the report and subsequent correspondence, it is clear that there were a multitude of factors frustrating Dale's progress. Most notably, in addition to the lack of resources, Dale complained of the challenges presented by no single department being responsible for science broadcasts.[52] After reading Dale's report, Mary Somerville suggested that the lack of resources, and the ill-thought-through remit given to Dale, were symptomatic of senior BBC figures' resistance to the committee's recommendation that scientific output required external oversight.[53] Whether this accusation was true or not, what was evident in Dale's report is that even with the right support and

[49] For more on the links between C. P. Snow and Jacob Bronowski, particularly the development of their humanist beliefs, and how they were defined in distinct divergence from prominent Christian humanists while they were both students at Cambridge University, see: Robert Westman, 'The "Two-Cultures" Question and the Historiography of Science in the Early Decades of the Salk Institute for Biological Studies', *Sartoniana* 32 (2019), http://www.sartonchair.ugent.be/file/293, 60–67.

[50] See Memorandum, "Suggestions for Scientific Adviser", January 1, 1950, and other documents related to the process of appointing Dale in folder R6/186-GAC, BBC-WA.

[51] Memorandum from George Barnes (Director of Television, BBC) to Controller of Television, January 29, 1951, R6/186-GAC, BBC-WA.

[52] Henry Dale, "Report to the BBC Governors", January 13, 1953, R6/186-GAC, BBC-WA.

[53] In reflecting on the report, Mary Somerville referred to the BBC's treatment of Dale as, "a somewhat unhappy day in BBC history". Memorandum from Mary Somerville (Controller, Talks Department) to (Director, Spoken Word), January 26, 1953, R6/186-GAC, BBC-WA.

a refined remit, by the early 1950s this role was not best undertaken by a senior scientist. However well connected Dale or any successor may have been the role needed someone who was an expert in communication and translation. Someone who was scientifically literate, but who also understood the process of broadcasting and in particular television's potential for science broadcasting.[54]

Internally, BBC staff used Dale's report to argue for more resources and coordination of science broadcasting. However, ultimately, they positioned Dale as out of step with audience demand, and juxtaposed his old-fashioned suggestions against the high level of in-house expertise already at the BBC.[55] In this wider institutional context, Clow's 1952 report on evolution can be seen as an example of BBC producers reinforcing their expertise and autonomy from organisations like the Royal Society and the British Association for the Advancement of Science. Clow's report contained not only advanced scientific content but also specific reflections and instructions related to the process of broadcasting. If written by an external scientist, the report would have read very differently. The report set out the parameters for broadcasters working on evolution. It was not a departure from previous approaches, but it simply codified the way the corporation had produced radio content on evolution to date. It helped to formalise the normative framing used for scientific content on evolution, centred on solid, but rapidly developing scientific understanding—modern, progressive and wonderful, but importantly still incomplete. Clow's report empowered broadcasters when depicting evolution to include both scientifically and epistemologically critical voices. However, such criticism was to be metered and sit within the bounds of sound science and philosophy; Clow made it clear that there was no space for positions that were anti-evolutionist.

Although Clow's report was on evolution, and thus contained points that were a specific rebuttal of anti-evolutionary positions, much of the text gave guidance relevant to any science broadcast that had the potential to be contentious for sections of the audience. The approach advocated in Clow's report, of showing the contingent but progressive nature of science and including carefully curated critical voices, had its roots in the

[54] For more on Dale's time as BBC Science Adviser and his report see: Jones, 'Clogging the Machinery'.

[55] See, for example, Memorandum from Mary Somerville (Controller, Talks Department) to Director, Spoken Word, January 26, 1953, R6/186-GAC, BBC-WA.

science and society format developed at the BBC in the 1930s. Driven by broadcasters' desire to make science relatable to a general audience, it was predominantly left-wing scientists and popularisers who had promoted the science and society approach in its original iteration.[56] In his letter to the Director-General that triggered the 1949 review on science broadcasting, Mark Oliphant echoed this point, stating, "I don't think scientists should always appear as Utopian idealists, as Marxists, or as amateur politicians".[57] The repeated contestation over the representation of science on radio and television in this period was not just about who had the expertise and authority to speak for science; rather it was one institutional manifestation of larger post-war political wrangling, where various parties were attempting to co-opt scientific endeavour to further a particular geopolitical vision of the future.[58]

While for the scientists' involved the broadcasting review on science was about alleviating fears emanating from the wartime co-opting of scientific breakthroughs, senior figures at the BBC saw the sub-committee's investigation in 1949 as a timely opportunity to explore the creative potential that television presented for communicating science. However, here again progress was thwarted somewhat by the appointment of Dale as the first Scientific Adviser. Like Oliphant, Dale wanted to emphasise scientific knowledge independent of application, and saw television as just like a public-facing science lecture, but with a significantly bigger audience.[59] Dale's unimaginative vision for television was out of step with that of the younger generation of BBC producers, who in the early 1950s—fuelled by a boom in the number of viewers (Fig. 3.1) and improving technological capabilities—were experimenting with new techniques and developing new genres of science broadcasting.[60]

One of those experimenting with broadcasting techniques and presentation styles in this period, albeit for radio, was Archibald Clow. In response to audience research reports, Clow had spent several years shifting the

[56] For more see Chap. 2 and Allan Jones, 'Mary Adams and the Producer's Role in Early BBC Science Broadcasts', *Public Understanding of Science* 21:8 (2012): 968–983, https://doi.org/10.1177/0963662511419450.

[57] Letter from Mark Oliphant to W. J. Haley (BBC Director-General), May 16, 1949, R6/34, BBC-WA. All BBC copyright content reproduced courtesy of the British Broadcasting Corporation. All rights reserved. As quoted in Jones, 'Clogging the Machinery', 437–438.

[58] David Edgerton, *Warfare State: Britain, 1920–1970* (Cambridge University Press, 2005), Chapter 8.

[59] Boon, *Films of Fact*, 190.

[60] Boon, 192.

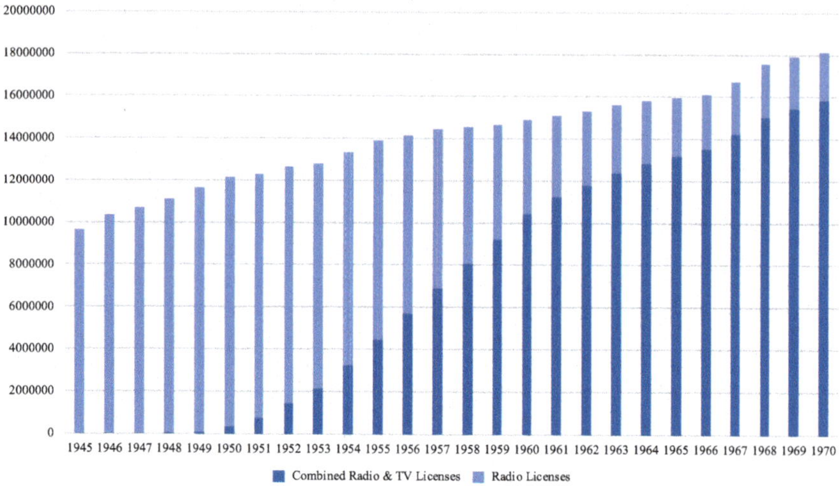

Fig. 3.1 Total radio and TV licenses by year in the UK. BBC Yearbooks and Handbooks, 1947–1970

format of the weekly Forces Educational Unit (FEU) radio show *Science and Everyday Life* from industrial scientists giving lectures to a magazine format, which interspersed scientists speaking with field recordings, narration and comment from other participants, such as journalists.[61] From 1948 onwards, the science writer and founding member of the Association of British Science Writers, A. W. Haslett narrated *Science and Everyday Life*, in Clow's words to "point [out] the moral and make the implications of the information doubly clear".[62] In 1952 Nan Clow, Archibald's wife, who had been brought in as an interim producer "to relieve him of chores"[63] ran an experiment assessing seven different presentation styles used on the series. Tasked with the "chore" of producing the 1952 series of *Science & Everyday Life*, Nan, a chemist and long-time collaborator of Archibald's, took the opportunity to run structured audience research

[61] Jared Robert Keller, 'A Scientific Impresario: Archie Clow, Science Communication and BBC Radio, 1945–1970' (Imperial College London, PhD Thesis, 2017), http://spiral.imperial.ac.uk/handle/10044/1/57504, 76–80.

[62] Archibald Clow to D.T., "Science in Forces Educational Broadcasts," January 19, 1946, R51/530, BBC-WA. All BBC copyright content reproduced courtesy of the British Broadcasting Corporation. All rights reserved. As quoted in Keller, 80.

[63] Memorandum from Mary Somerville (Controller, Talks Department) to Director, Spoken Word, June 9, 1952, R6/186-GAC, BBC-WA. All BBC copyright content reproduced courtesy of the British Broadcasting Corporation. All rights reserved.

alongside the series as it was broadcast. Finding that different presentation styles worked best for differing levels of education, Nan recommended for broadcasts aimed at educating the armed forces that "some form of dilution of the scientific material" was required.[64] In keeping with experimental approaches developed by the FEU, and Archibald's earlier development of the series, Nan found that this was best achieved by using professional science journalists and broadcasters to narrate and analyse the content contributed by practicing scientists.[65]

While most of the experimentation with science formats for radio in the post-war years was in departments with an educational remit, such as FEU, in the early 1950s all science broadcasting on television was still in an experimental phase.[66] Thus, nearly all departments and units who produced science content for television in this period were involved in trials and experimentation with new approaches.[67]

During the 1950s, as transmission coverage across the UK increased, so too did television ownership, growing rapidly from under 400,000 licensed sets at the start of the decade to over nine million by the end of 1959 (Fig. 3.1).[68] Alongside this fast growth in audience, rapid technological advancements enabled the development of new approaches, styles and formats. Most notably, in 1952 the BBC moved into Lime Grove television studios and in 1953 began its protracted move to purpose-built television studios in Shepherd's Bush, west London. As the decade progressed, the capacity for outside broadcasts grew, and the capability for live multi-location link-ups was established. The passing of the Television Act in 1954, and the subsequent rolling out of regional commercial television in the UK, via the launch of Independent Television (ITV) in 1955, further drove an emphasis on innovation in BBC television.[69]

[64] Nan Louise Clow, "Report on F.E.B.s – Science and Everyday Life," June 19, 1952, 1, R51/530, BBC-WA, 2. All BBC copyright content reproduced courtesy of the British Broadcasting Corporation. All rights reserved. As quoted in Keller, 'A Scientific Impresario', 81.

[65] For a more detailed account of this episode and more on the working relationship between Archie and Nan, see Keller.

[66] For more on educational broadcasting see Chap. 4.

[67] By the mid-1950s Talks, Drama, Children's Programmes, Features, Outside Broadcasts and Films were all contributing science content for television. *The British Broadcasting Corporation Yearbook* (HMSO, 1956).

[68] By the end of 1956, a network of 18 transmitters meant that an estimated 98% of the British public could receive television signals (*The British Broadcasting Corporation Yearbook*, HMSO, 1956, 27). For more on the growth and development of television audiences in the period see: Briggs, *The History of Broadcasting in the United Kingdom*, 220–244.

[69] Briggs, 805–919.

For the cohort of ambitious writers, directors and producers—the first to have been trained specifically for television, rather than film or radio—this landscape provided a fertile backdrop for the creation of many new and enduring television formats. While the earliest science on television had directly broadcast, or simply replicated, popular public lectures and science demonstrations,[70] by the early 1950s science on television was expanding into areas including panel shows, such as *Animal, Mineral, Vegetable* (1952–1959) and *The Brains Trust*, which made the switch from BBC radio to television in 1955.[71] Early successes with what were coined documentary-dramas, such as *Matters of Life and Death* (1949–1951), a 14-part series covering different aspects of modern medicine, began to push the format beyond the simple replication of film and radio documentaries.[72] *Matters of Life and Death* caused some controversy with its frank and visceral depictions of illness and medical issues, with some scenes cut for being too explicit, while reviewers praised those that made the final broadcast for their frank and refreshing approach.[73]

From our contemporary digital multimedia landscape, the shifts undertaken as non-fiction science television formats emerged from more

[70] Broadcasters were still widely using these approaches in the early 1950s, for examples from this period see: *From Small Beginnings*, presented by Professor Edward Andrade, produced by George Noordhof, Jan–March 1953, BBC Television; and from September 1951 onwards the annual broadcasting of the British Association for the Advancement of Science's Presidential Addresses.

[71] In *Animal, Mineral, Vegetable* a panel of relevant experts, including archaeologists and natural historians were tasked with identifying unusual and interesting artefacts and objects that had been brought into the studio. The interested reader can watch surviving clips from the series online at https://www.bbc.co.uk/programmes/p017bdl3 *(last accessed 18/06/2021)*.

[72] The series was the first with a scientific focus to be singled out for commendation in the television section of the BBC Yearbook, and although its explicit medical imagery was considered unseemly by some commentators, newspaper reviewers lauded its producer Andrew Miller Jones, and the production techniques he used. See: Our Radio Critic. "Medical Programmes", *The Manchester Guardian*, April 11, 1951: 3; J. Stubbs-Walker, "All Came Right in the End", *Daily Mail*, April 10, 1951: 2; *The British Broadcasting Corporation Yearbook* (HMSO, 1952), 107; and *Matters of Life and Death*, episodes 1–14, produced by Andrew Miller Jones, October 1949–April 1951, BBC Television.

[73] Episode four "Modern Surgery" was supposed to feature footage of appendicitis surgery, but upon viewing the film, and following pressure from the patient's family, BBC officials deemed the film not suitable for public exhibition. Anon. "Operation Girl Makes TV Protest", *Courier and advertiser* [Dundee, Scotland], February 17, 1950: 3; and J. Stubbs Walker, "Last Night's Television", *Daily Mail*, March 21, 1950: 4.

traditional film and radio documentary approaches may seem minor and even inconsequential. However, for audiences experiencing television first-hand as a new live phenomenon, the differences and benefits were obvious and apparent. As one radio critic, reviewing *Matters of Life and Death*, reflected:

> Television has indeed come a long way from its early and laborious stages to this smooth and apparently effortless, though in fact complicated, production, in which eminent specialists and professors in turn took up the story of one hundred years of medicine and surgery. Illustrations of every kind—from slides under the microscope to models of the mosquito, from early pictures of primitive surgery to the actual sight of the modern operating theatre and the methods of anaesthesia—were dovetailed into one steady sequence which had the merits of a good documentary film with the addition that most of what one saw was actually happening. Andrew Miller Jones, the producer, must be congratulated on this programme.[74]

While some of the earliest contestations over science on television had been between live broadcasts and the older pre-television format of documentary films, by the early 1950s, competition for science's place in the schedule was centred on two live formats native to television; studio-based and location-based.[75] Building on the early successes produced by Andrew Miller Jones, from 1955 onwards former science teacher James McCloy became the most prominent television science producer at the BBC.[76] McCloy led on making a wide range of shows—including panel shows, such as *A Question of Science* (1956–1957), which borrowed from the format of *The Brains Trust* and regularly featured the biologist Professor Peter Medawar (1915–1987);[77] and science magazines, such as *Science is News* (1958–1959) a live review of new scientific developments, which replaced the earlier film and newsreel approach of *Science Review*

[74] Our Radio Critic. "Medical Programmes." *The Manchester Guardian*, April 11, 1951: 3.

[75] Boon, *Films of Fact*, 210.

[76] "James McCloy." *The Times*, August 19, 2002: 7.

[77] Professor Peter Medawar was a pioneer of transplant immunology, who was Jodrell Professor of zoology and comparative anatomy at University College London (1951–1962). Alongside his groundbreaking research, for which he shared the Nobel Prize in 1960, he was also a prolific popular science writer and broadcaster, commenting on a wide range of subjects beyond biology, from religion to career advice for young scientists.

(1952–1956).[78] Both of these popular formats helped to establish science programming as a central component of the television schedule in the UK.

Alongside McCloy's work in the Television Talks department, during the 1950s natural history, which had previously been the preserve of the Films department, began to appear in studio-based formats. In 1952, attempting to redress a perceived imbalance towards the arts, Mary Adams recruited two young trainee producers with scientific backgrounds to Television Talks: the first was James McCloy and the second was David Attenborough. After cutting his teeth on a range of current affairs shows, Attenborough had set about producing natural history programmes in Adams' science and society mould. After collaborating on a one-off special on the prehistoric fish *The Coelacanth* (1953), Attenborough pitched a series on the evolution of animal colouration to Julian Huxley, a particular challenge remembering BBC Television was still a black and white only service.[79] *Animal Patterns* (1953) was a three-part series, which focused respectively on camouflage, warning signals and courtship displays. The mini-series, broadcast live, featured Huxley discussing film sequences, photographs, models, stuffed specimens and live animals, and made use of the new camera technologies available in the Lime Grove studios.[80] While Huxley's input ensured the mini-series was scientifically sound and focused on animal physiology in relation to evolution, his competence, as a studio presenter for this new televisual approach, was less convincing. As one reviewer summarised:

> The information is, of course, authoritative and abundant, and anyone who listens closely to Dr Huxley will learn a great deal in thirty crowded minutes. But the speaker's style can hardly be called encouraging; the sidelong glances at his notes, the dry, detached air with which he prods a creature to make it show its tricks, seem more in keeping with the lecture-theatre than the television screen.[81]

[78] Boon, 210–214.

[79] See TV Art 1, Julian Huxley, Sir, File 1, 1937–1962 (BBC-WA), which includes extensive handwritten notes, possibly in David Attenborough's hand, on *The Coelacanth*, dated January 13, 1953; and Attenborough's original pitch and letters to Huxley for what became the mini-series *Animal Patterns* (March–April, 1953).

[80] Jean-Baptiste Gouyon, *BBC Wildlife Documentaries in the Age of Attenborough* (Springer Nature, 2019), 49–53.

[81] C. A. Lejenue, *The Observer*, July 12, 1953: 6.

The reviewer went on to negatively contrast Huxley's on-screen demeanour against that of the most popular natural history broadcaster of the period, George Cansdale.[82] Popular with both children and general audiences, Cansdale was not an academic, but a former Forestry Officer and Superintendent of the London Zoological Society. He had found fame on both radio and television from the late 1940s onwards with his more informal, but still educational, approach to natural history, in shows such as *Looking at Animals* (BBC TV, 1951–1954). When producing *Animal Patterns*, Attenborough attempted to mediate some of Huxley's most highbrow tendencies; successfully petitioning Mary Adams to remove Huxley's proposed fourth episode, which was to feature the evolutionary biologist Niko Tinberg, and was "rather more difficult, intellectually".[83] Even Huxley, with his wealth of experience in communicating science to popular audiences, struggled with the live and visual elements that the transition to television demanded. Attenborough, with his dedicated training in television, understood the importance of a clear communicative broadcaster, who could deal with the live and visual aspects. However, as a natural sciences graduate, he also believed that the right scientific expert could exploit the didactic potential of television. Thus, following *Animal Patterns* he sought a way forward that could combine the entertainment value of Cansdale, with the synthetic scientific overview of Huxley.

Keen to create a niche for himself as the go-to producer for natural history in the Talks department, Attenborough followed *Animal Patterns* with a pioneering new show, *Zoo Quest* (1954–1962). Made initially in partnership with the London Zoological Society, *Zoo Quest* followed Attenborough as he accompanied expeditions to far-off climes gathering new zoo specimens.[84] The animals were transported back to the UK, and

[82] In this area of more entertainment centred natural history programming, the BBC were also competing with ITV, in particular their long-running success *Zoo Time* (1956–1968), presented until 1959 by the charismatic zoologist and science populariser Desmond Morris. Gouyon, *BBC Wildlife Documentaries in the Age of Attenborough*, 64–67 & 82–87.

[83] David Attenborough to Mary Adams (Head of Talks Television), March 10, 1953, BBC-WA, T32/330/4. All BBC copyright content reproduced courtesy of the British Broadcasting Corporation. All rights reserved. As quoted in Gouyon, *BBC Wildlife Documentaries in the Age of Attenborough*, 50.

[84] For more on the *Zoo Quest* expeditions see: David Attenborough, *The Zoo Quest Expeditions: Travels in Guyana, Indonesia, and Paraguay* (James Clarke & Co., 1980). Notably this autobiographical travelogue contains very little discussion of *Zoo Quest* the television show, barely mentioning the studio element and focusing on the adventures Attenborough had while travelling the world filming and collecting specimens.

brought into the television studio to be discussed by relevant experts, with footage from the expeditions interspersed throughout the broadcast.

With *Zoo Quest* Attenborough managed to find his middle way, combining intellectual rigour with entertainment of interest to the whole family. Guided by Mary Adams, storytelling was a central feature of these series, as Attenborough successfully married the travelogue exploration tradition of early documentary film[85] with Cansdale's approach of bringing the animals up close and centre stage in the studio (Fig. 3.2).[86] With continued input from relevant scientific experts, but this time mediated via the right communicator—Attenborough himself—the show reached a popular primetime audience, while also managing to talk in simple language about evolutionary adaptations, phylogenetic relationships, and other relevant issues, such as habitat loss.[87]

From 1957 onwards, the Television Talks department faced renewed internal competition, as Aubrey Singer became Head of Outside Broadcasts and reinvigorated the department's science content. Singer's first major science broadcast in this role, *The Restless Sphere* (1957), typifies the scale and ambition he brought to outside broadcasts on science. Aired on June 30, 1957, with narration provided by the Duke of Edinburgh Prince Philip, this science special detailed the year-long activities undertaken by scientists around the world as part of the 1957 International Geophysical Year.[88] As well as including film recorded in 12 countries, *The Restless Sphere* featured live outside broadcast links to four locations across Europe. The most remarkable were the very first live broadcasts from the crater of Mount Vesuvius and from the Aletsch Glacier in Switzerland, some

[85] For more on the early pioneers of this tradition see: Erik Barnouw, *Documentary: A History of the Non-Fiction Film* (Oxford University Press, 1993), 33–51.

[86] Gouyon, 52–68. Readers in the UK can watch clips from the *Zoo Quest* series online (https://www.bbc.co.uk/programmes/p03qxfsg/clips) and all of the episodes of the final series of *Zoo Quest* from 1961, which followed an expedition to Madagascar: https://www.bbc.co.uk/iplayer/episode/p00db24v/zoo-quest-zoo-quest-to-madagascar-episode-1 *(last accessed 17/06/2021).*

[87] For an example of Attenborough's skill in introducing information about a species' evolutionary history without using complex language see the closing section of *Zoo Quest to Madagascar: Episode 4* (from 26 minutes onwards). *Zoo Quest to Madagascar: Episode 4*, presented and produced by David Attenborough. June 23, 1961. BBC Television. Available online to readers based in the UK: https://www.bbc.co.uk/iplayer/episode/p00db25t/zoo-quest-zoo-quest-to-madagascar-episode-4 *(last accessed 17/06/2021).*

[88] Boon, *Films of Fact*, 214–216.

Fig. 3.2 Prince Charles and Princess Anne in the studio along with Attenborough's pet cockatoo 1958. BBC © BBC Photo Library

3500 metres above sea level.[89] The historian of science Tim Boon has described the show as a "daring deployment of outside broadcast technology", which "represented a new breed of television science programme".[90] Indeed, the success of *The Restless Sphere* led to Singer's breakthrough science series, *Eye on Research* (1957–1961), which took the live outside

[89] Interested readers based in the UK can watch *The Restless Sphere* online here: https://www.bbc.co.uk/archive/the-restless-sphere%2D%2Dthe-story-of-the-international-geophysical-year/zbqf7nb *(last accessed 17/06/21).*

[90] Timothy Boon, "The Restless Sphere, Outside Broadcast and the Invention of Science Television," *British Quality, American Chaos? Trans-national discourses and interactions in the history of British and North American broadcasting, circa 1922–1962.* AHRC Early Broadcasting History Network Conference, University of Wales Study Centre, Gregynog, Newtown, Mid-Wales, July 20–22, 2011.

broadcast cameras and its journalist-styled presenter Robert Reid into various laboratories and scientific establishments.[91]

All these emerging science formats required large production teams and budgets, and, with limited camera equipment and studio space, required extensive planning. For most research and production staff at the BBC, the trials and experiments on radio and television were pragmatic attempts to increase viewing figures. However, for the handful of producers concerned with ensuring science content had an enduring place in the schedule, the format developments resulting from these trials had to be situated within a more systematic vision for the role and purpose of science broadcasting in the UK. Despite the influx of these young producers and their creative new ideas, the systematic framework for communicating science, in which all of their work in this period was situated, was philosophically grounded in Huxley's scientific humanism.[92] Clow's time in Aberdeen with Lancelot Hogben clearly influenced his report on evolution, while Miller, McCloy and Attenborough were all working under the mentorship of Mary Adams, who during the 1930s had been the first to apply this progressive vision of humanity to broadcast media. As well as Attenborough's new studio-based approach to natural history, other traditional approaches to natural history at the BBC, which centred on field recordings of species, were also firmly rooted in this worldview. One element of post-war discussions to better society was on the "social usefulness" of natural history, a key driver in establishing the, now world-renowned, BBC Natural History Unit in Bristol in July 1957.[93]

Of these young, predominantly male producers, Aubrey Singer had the least scientific background, and perhaps as a result, was often the most

[91] For more on the similarly influential US television coverage of the IGY, see Fae L. Korsmo, 'Shaping Up Planet Earth: The International Geophysical Year (1957–1958) and Communicating Science Through Print and Film Media', *Science Communication* 26:2 (2004): 162–187, https://doi.org/10.1177/1075547004270076.

[92] This is not the only area where Julian Huxley has been considered a bridge figure between different generations and the associated development of ideas. Weindling (2012) outlines how Huxley acted a crucial figure bridging the divide between old eugenics and a later iteration of eugenic ideas informed by molecular biology. Paul Weindling, '"Julian Huxley and the Continuity of Eugenics in Twentieth-Century Britain"', *Journal of Modern European History*, 10:4 (2012): 480–499, https://doi.org/10.17104/1611-8944_2012_4.

[93] Gouyon, *BBC Wildlife Documentaries in the Age of Attenborough*, 28–31. For more on the history of the Natural History Unit see: Gail Davies, "Networks of Nature: Stories of Natural History Film-Making from the BBC" (University College London, PhD Thesis, 1998).

pragmatic in his approach. As with the other successful popularisers of science from the period, including Huxley and Clow, he understood the important role of narrative in communicating science via broadcast media. Like Nesta Pain before him (Chap. 2), Singer prioritised the story being told, understanding that even for a public service broadcaster; science broadcasting must first and foremost be entertainment. As he summarised:

> The televising of science is a *process of television*, subject to principles of programme structure and the demands of dramatic form. Therefore in taking programme decisions, priority must be given to the medium rather than scientific pedantry.[94]

Despite this pragmatic grounding, Singer's vision for science on television was still completely situated in a progressive, technocratic vision of a scientific society. Reflecting on the role of television in society in 1965 Singer stated:

> Perhaps we've not lost our vision of Utopia. Perhaps it's changing. As man changes. Perhaps what we're all a party to is a struggle between *man* and *mankind*: the point of evolutionary decision between Homo sapiens and (dare I coin the word) Homo cyberneticus.[95]

Singer's reference to man's evolutionary future, his nod towards what we might today call transhumanism, and the language he used throughout this talk at the "World Congress on New Challenges to Human Communication" was influenced by Julian Huxley. Here again we see the wide reach of Huxley's writing and broadcasts on scientific humanism. Thus, while in the post-war years there was no single department at the BBC responsible for making science broadcasts, those responsible for producing the majority of science output, whether directly or indirectly, were all influenced by Huxley and his vision for the progressive role of science in society.

[94] Aubrey Singer, "Science Broadcasting", *BBC Lunch-Time Lectures*, series 4, no. 1–6 (British Broadcasting Corporation), as quoted and discussed in Boon, *Films of Fact*, 232; and Roger Silverstone, *Framing Science: The Making of a BBC Documentary* (BFI Publishing, 1985), 160–180. All BBC copyright content reproduced courtesy of the British Broadcasting Corporation. All rights reserved.

[95] Aubrey Singer was addressing the audience at the Vision 65: World Congress on New Challenges to Human Communication, held in Carbondale, Illinois October 21–23, 1965. Aubrey Singer, "Television: Window on Culture or Reflection in the Glass?" *The American Scholar*, 35:2 (1966): 303–309.

Despite the development of new science formats for television, Clow's report—which called for more open and in-depth coverage of evolution as early as 1952—and the relative popularity of shows that tangentially covered evolutionary themes (e.g. *Zoo Quest*), it wasn't until 1958 that the subject received its own dedicated television treatment.

Evolution on the Small Screen

The first show on British television about evolution, *Five Hundred Million Years*, was a series of six episodes broadcast in April and May 1958, to coincide with the 100th anniversary of the joint reading at the Linnaean Society of Charles Darwin's and Alfred Russel Wallace's essays on evolution.[96] The series, a combination of talking heads and panel debates interspersed with relevant stock footage, aimed to show viewers "how evolution looks today a hundred years after Darwin".[97] After considering Darwin's original work, natural selection and how recent discoveries on heredity were being incorporated into the latest scientific understandings, the show dedicated the final three episodes to man's evolution. In the final episode, on "The Nature of Man", three eminent biologists and the Anglican Reverend J. S. Habgood debated whether evolutionary theory contradicted religion.[98] Although there was some significant divergence among the three biologists—Julian Huxley, Professor of Anatomy, J. Z. Young (1907–1997), and anatomist and Roman Catholic, Bernard Towers (1922–2001)—as the chair, the progressive educationalist Sir John Wolfenden, stated at the outset, all agreed "about the facts and the principles of organic evolution".[99] The discussion that followed focused on the implications of evolutionary theory for subjects such as *Homo sapiens*' uniqueness, man's soul, the immaterial and moral relativism. These are all subjects that will be recognised by those familiar with historical debates

[96] *Five Hundred Million Years*. Produced by James McCloy. BBC, April–May 1958. Transcripts held on microfilm at BBC-WA.

[97] Five Hundred Million Years: 1: Did it Happen?, *Radio Times*, April 18, 1958: 15.

[98] Only 31 at the time of filming, Habgood would go on to become the Bishop of Durham (1973–1983) and the Archbishop of York (1983–1995). He continued to be interested in science and religion throughout his career; most notable in this regard is his publication *Truths in Tension: New Perspectives on Religion and Science* (Holt, Rinehart and Winston, 1964).

[99] *Five Hundred Million Years*, "The Nature of Man," produced by James McCloy, aired May 27, 1958, on BBC TV. Transcript held on microfilm at BBC-WA.

about evolution, particularly those generated by Darwin's publications in the second half of the nineteenth century.[100]

While this series was the first television show about evolution in the UK, by 1958 the BBC had already aired over 80 individual episodes on the subject via its 2 radio stations. Many of the scientists who featured on *Five Hundred Million Years*, including Professor Peter Medawar, Professor J. Z. Young and Sir Julian Huxley, had all featured on previous radio shows about evolution on the BBC. Commercial specialists in scientific and educational films, the Macqueen Film Organisation, provided most of the film sequences used during the series.[101] This external partnership was an early nod towards the kind of collaboration that would later become a central component of the co-production approach of big-budget science documentaries in the 1970s (Chap. 6).

Five Hundred Million Years received positive reviews from the few critics who covered it,[102] and the fact an amended version aired in the children's hour slot later that summer suggests those at the BBC also considered it a success.[103] Although it borrowed heavily from the format of prior radio shows on evolution, the approach used on *Five Hundred Million Years*, both with regard to the use of talking heads and stock footage, and the structure of the content itself, influenced subsequent television science series. For example, the way the show's presenter, William van

[100] For more on some of these debates in a nineteenth-century context, see: Bernard Lightman and Bennett Zon, *Evolution and Victorian Culture* (Cambridge University Press, 2014); James R. Moore, *The Post-Darwinian Controversies: A Study of the Protestant Struggle to Come to Terms with Darwin in Great Britain and America, 1870–1900* (Cambridge University Press, 1979); and Ian Hesketh, *Of Apes and Ancestors: Evolution, Christianity, and the Oxford Debate* (University of Toronto Press, 2009).

[101] The Macqueen Film Organisation was an independent film production company founded by Norman Macqueen. In addition to editing and producing film for BBC science programmes, Macqueen chaired ITV's *The Scientist Replies*, and the company made thousands of films for a wide range of customers including universities, industrial companies, and learned societies such as the Royal Institution. "Mr. Norman Macqueen", *The Times*, October 9, 1959, 15; Advert, *The New Scientist*, April 30, 1959, 946; and "Bragg to Norman S. Macqueen, Macqueen Film Organisation", *Royal Institution of Great Britain* W.L. BRAGG/27C/36. For a partial filmography of the Macqueen Film Organisation's productions see: https://www2.bfi.org.uk/films-tv-people/4ce2b944037b3 *(last accessed 17/06/2021).*

[102] See, for example, Peter Black, "Peter Black's Teleview," *Daily Mail*, May 28, 1958, 10; and Maurice Wiggin, "It's Later Than They Think," *Sunday Times*, May 18, 1958, 12.

[103] British Broadcasting Corporation, *BBC Yearbook 1960*, 55.

Essen "fitted each little lecture to the next", today an unremarkable standard practice of television documentaries, was commended at the time.[104] The emergence of a relatively standardised way of constructing shows about evolution can be considered part of a wider process of standardisation, which saw the genre (and sub-genres) of television science solidifying during the period.

While scientists and their representative bodies continued to raise issues surrounding the best way to communicate science via television,[105] ultimately audience figures and feedback were the primary drivers for the development of science content. As the Director of BBC Television, Kenneth Adam remarked during deliberations, "[i]t comes down in the long run to whether you can stand the sight of Dr Bronowski on the screen. This is entirely a subjective matter".[106] The explosion in television's popularity, and the BBC's ever-expanding output aimed at increasingly diverse target audiences, ensured that science, and in particular evolutionary content, continued to be popular. As we shall see in the coming chapters, it was not just in traditional science broadcasts that evolutionary themes found a home, frequently appearing across a plethora of emerging formats from educational re-enactments through to science fiction series.

Conclusion

In this chapter we have seen how in the decade and a half following the Second World War, as part of wider reflections on the role of science in shaping the future of society, the scope of radio broadcasts that included content on evolution continued to expand. Whether shaped by religious belief, as in the broadcasts of Rev. Charles E. Raven, or by secular humanist ideals, as in the radio broadcasts of Jacob Bronowski, the narrative this content promoted was within an evolutionary framework. In these broadcasts, we can hear the lasting influence of the small group of left-wing

[104] Anthony Curtis, "Critic on the Hearth," *The Listener*, May 1, 1958, 748.

[105] In 1958 the Royal Society and the British Association for the Advancement of Science renewed their calls for a dedicated department for the broadcast of science, and throughout the early 1960s contestation over science content continued. See TV Policy: Science and Technology, 1958–1968, T16/623 BBC-WA; and Boon, *Films of Fact* 221–226.

[106] Kenneth Adams in response to statement by Stafford Beers, Copy of Minutes of BBC General Advisory Council, 1959, T16/623 BBC-WA. All BBC copyright content reproduced courtesy of the British Broadcasting Corporation. All rights reserved.

intellectuals, so briefly aligned in the late 1920s and 1930s, not just through Julian Huxley and his continuing popular broadcasts, but also through the influence of Lancelot Hogben on Archibald Clow, and in the work of Mary Adams.

This chapter has begun to untangle how this group of intellectuals fundamentally influenced how we talk about science in popular culture. Through the launch of television, Clow's expression of an editorial position on evolution, and early broadcasts on the subject, we have also seen how their mode of science broadcasting, informed by scientific humanism, effected the actual structure and format of emergent science broadcasting genres. As we will see in the following chapters, these early forays would go on to have a lasting effect on the dominant narratives and frames used in future broadcasts on evolution and for the natural sciences more generally.

Much of the broadcast content covered in this chapter, even *Five Hundred Million Years*, did not provide an in-depth treatment of contemporary evolutionary science. More frequently, these broadcasts painted grand-historical narratives or reflected on the implications of evolutionary science for modern society. While the BBC did produce more in-depth science content on the subject, such as *The Process of Evolution* (1951), this was often less popular or targeted at specific audiences, such as school students (Chap. 4). In his report on evolution, Archibald Clow complained that although the science had moved on considerably from "initial Darwinism", anti-evolutionists were still critiquing an outdated strawman version of evolution.[107] However, in response to Clow's draft Henry Dale stated, "I am baffled by the thought of making anything but the barest and most dogmatic out-lines of such matters intelligible over the air".[108] This tension between imparting the latest (complex) scientific knowledge and the intelligibility and popular appeal of the narrative presented has remained central to broadcasting on evolutionary science. Whether science broadcasting should limit itself to just imparting the latest scientific knowledge or whether it should be confident in telling stories and creating idealised narratives remains contested and contingent even today.

[107] Clow, "Report on Evolution," 10.

[108] Letter from Sir Henry Dale (BBC Science Advisor) to Archibald Clow (Science Producer, Talks), 18 March, 1952, 93HD 29.2, Royal Society Archive.

References

Advert, *The New Scientist*, April 30, 1959, 946.

Agar, Jon. *Science in the 20th Century and Beyond*. Polity, 2012.

Alvargonzález, David. 'Is the History of Science Essentially Whiggish?' *History of Science*51:1(2013):85–99.https://doi.org/10.1177/007327531305100104.

Anon. "Operation Girl Makes TV Protest", *Courier and Advertiser* [Dundee, Scotland], February 17, 1950: 3.

Artifex. "Science and the Christian Man", *The Manchester Guardian*, March 24, 1952.

Attenborough, David. *The Zoo Quest Expeditions: Travels in Guyana, Indonesia, and Paraguay*. James Clarke & Co., 1980.

Barnouw, Erik. *Documentary: A History of the Non-Fiction Film*. Oxford University Press, 1993.

BBC. *The British Broadcasting Corporation Yearbook*. Her Majesty's Stationery Office, 1952.

———. *The British Broadcasting Corporation Yearbook*. Her Majesty's Stationery Office, 1956.

———. *The British Broadcasting Corporation Yearbook*. Her Majesty's Stationery Office, 1960.

Black, Peter. "Peter Black's Teleview," *Daily Mail*, May 28, 1958, 10.

Boon, Timothy. *Films of Fact: A History of Science in Documentary Films and Television*. Wallflower Press, 2008.

———. "The Restless Sphere, Outside Broadcast and the Invention of Science Television," *British Quality, American Chaos? Trans-national discourses and interactions in the history of British and North American broadcasting, circa 1922–1962*. AHRC Early Broadcasting History Network Conference, University of Wales Study Centre, Gregynog, Newtown, July 20–22, 2011.

Bowler, Peter J. *Reconciling Science and Religion*. University of Chicago Press, 2010.

Briggs, Asa. *The History of Broadcasting in the United Kingdom: Volume IV: Sound and Vision*. Oxford University Press, 1995.

Bronowski, Jacob. *The Poet's Defence*, Cambridge University Press, 1939.

———. *The Effects of the Atomic Bombs at Hiroshima and Nagasaki: Report of the British Mission to Japan*. His Majesty's Stationery Office, 1946.

———. *The Common Sense of Science*. Faber & Faber, 2011.

Bud, Robert. '"The Spark Gap Is Mightier than the Pen": The Promotion of an Ideology of Science in the Early 1930s'. *Journal of Political Ideologies*, 22:2 (2017), 169–181. https://doi.org/10.1080/13569317.2017.1298548.

Curtis, Anthony. "Critic on the Hearth," *The Listener*, May 1, 1958, 748.

Davies, Gail. "Networks of Nature: Stories of Natural History Film-Making from the BBC" (University College London, PhD Thesis, 1998).

Desmarais, Ralph. 'Jacob Bronowski: A Humanist Intellectual for an Atomic Age, 1946–1956'. *The British Journal for the History of Science* 45:4 (2012): 573–589. https://doi.org/10.1017/S0007087412001069.

Desmarais, Ralph John. 'Science, Scientific Intellectuals, and British Culture in the Early Atomic Age: A Case Study of George Orwell, Jacob Bronowski, P.M.S. Blackett and J.G. Crowther', (Imperial College London, PhD Thesis, 2010). http://spiral.imperial.ac.uk/handle/10044/1/5646.

Dewar, Douglas. "Letters: Broadcasting," *The Daily Sketch*, May 22, 1944.

Edgerton, David. *Warfare State: Britain, 1920–1970*. Cambridge University Press, 2005.

"Five Hundred Million Years: 1: Did it Happen?", *Radio Times*, April 18, 1958.

Gouyon, Jean-Baptiste. *BBC Wildlife Documentaries in the Age of Attenborough*. Springer Nature, 2019.

Habgood, J. S. *Truths in Tension: New Perspectives on Religion and Science*. Holt, Rinehart and Winston, 1964.

Herring, Emily. '"Great Is Darwin and Bergson His Poet": Julian Huxley's Other Evolutionary Synthesis'. *Annals of Science* 75:1 (2018): 40–54. https://doi.org/10.1080/00033790.2017.1407442.

Hesketh, Ian. *Of Apes and Ancestors: Evolution, Christianity, and the Oxford Debate*. University of Toronto Press, 2009.

Hopwood, Nick. *Haeckel's Embryos: Images, Evolution, and Fraud*. University of Chicago Press, 2015.

Huxley, Julian. *Evolution, the Modern Synthesis*. Allen & Unwin, 1942.

Huxley, Julian and Gavin De Beer. *The Elements of Experimental Embryology*. Cambridge University Press, 1934.

"James McCloy." *The Times*, August 19, 2002: 7.

Jones, Allan. 'Clogging the Machinery: The BBC's Experiment in Science Coordination, 1949–1953'. *Media History* 19:4 (2013): 436–449. https://doi.org/10.1080/13688804.2013.844892.

———. 'Mary Adams and the Producer's Role in Early BBC Science Broadcasts'. *Public Understanding of Science* 21:8 (2012): 968–983. https://doi.org/10.1177/0963662511419450.

Keller, Jared Robert. 'A Scientific Impresario: Archie Clow, Science Communication and BBC Radio, 1945–1970' (Imperial College London, PhD Thesis, 2017). http://spiral.imperial.ac.uk/handle/10044/1/57504.

Korsmo, Fae L. 'Shaping Up Planet Earth: The International Geophysical Year (1957–1958) and Communicating Science Through Print and Film Media'. *Science Communication* 26:2 (2004): 162–187. https://doi.org/10.1177/1075547004270076.

Lejenue, C. A. *The Observer*, July 12, 1953: 6.

Lightman, Bernard and Bennett Zon. *Evolution and Victorian Culture*. Cambridge University Press, 2014.

Lustig, Abigail, Robert J. Richards, and Michael Ruse. *Darwinian Heresies*. Cambridge University Press, 2004.
Moore, James R. *The Post-Darwinian Controversies: A Study of the Protestant Struggle to Come to Terms with Darwin in Great Britain and America, 1870–1900*. Cambridge University Press, 1979.
"Mr. Norman Macqueen", *The Times*, October 9, 1959, 15.
Numbers, Ronald L. *The Creationists: From Scientific Creationism to Intelligent Design*. Expanded ed edition. Harvard University Press, 2006.
Our Radio Critic. "Medical Programmes", *The Manchester Guardian*, April 11, 1951: 3.
Raven, Charles Earle. *Science and the Christian Man*. SCM Press, 1952.
Sandefur, Timothy. *The Ascent of Jacob Bronowski: The Life and Ideas of a Popular Science Icon*. Prometheus Books, 2019.
Silverstone, Roger. *Framing Science: The Making of a BBC Documentary*. BFI Publishing, 1985.
Singer, Aubrey. "Television: Window on Culture or Reflection in the Glass?" *The American Scholar*, 35:2 (1966): 303–309.
Smocovitis, Vassiliki Betty. *Unifying Biology: The Evolutionary Synthesis and Evolutionary Biology*. Princeton University Press, 1996.
Stubbs Walker, J. "Last Night's Television", *Daily Mail*, March 21, 1950: 4.
———. "All Came Right in the End", *Daily Mail*, April 10, 1951: 2.
Weindling, Paul. '"Julian Huxley and the Continuity of Eugenics in Twentieth-Century Britain"'. *Journal of Modern European History* 10:4 (2012): 480–499. https://doi.org/10.17104/1611-8944_2012_4.
Westman, Robert. 'The "Two-Cultures" Question and the Historiography of Science in the Early Decades of the Salk Institute for Biological Studies'. *Sartoniana* 32 (2019). http://www.sartonchair.ugent.be/file/293.
Wiggin, Maurice. "It's Later Than They Think," *Sunday Times*, May 18, 1958, 12.

CHAPTER 4

The Most Malleable of Minds: Evolution in Educational Broadcasting

> That this Public Meeting assembled in the Town Hall, Birmingham, on September 4th, 1944, expresses deep concern over the attitude of the B.B.C. in allowing the theory of Evolution to be broadcast to Schools in the series of talks "How Things Began". This meeting is of the opinion that such teaching threatens to undermine the faith of young people in the inspired Holy Scriptures and urges the B.B.C. to give an equal opportunity to speakers who accept the Biblical account of Creation. Wilfrid Phillips, Bible Testimony Fellowship, 1944.[1]

In September 1944, the BBC Director-General received a letter of complaint on behalf of 1000 people who supported the above resolution against radio broadcasts on evolution aimed at school children. The Birmingham and District Bible Testimony Fellowship created and circulated the resolution in response to the long-running Schools series *How Things Began* (1941–1968), which aimed to teach children about natural and human history. The episodes that triggered this specific complaint were "What is a Fossil", "Why Fossils are Found on Land" and "How Living Things Change", all written by the journalist and radio presenter

[1] Letter from Wilfrid Phillips (Bible Testimony Fellowship) to the BBC Director-General, September 15, 1944. BBC-WA R41/51/1.

A. Hall, *Evolution on British Television and Radio*, Palgrave Studies in Science and Popular Culture,
https://doi.org/10.1007/978-3-030-83043-4_4

Honor Wyatt.[2] Alongside Honor Wyatt, who wrote the natural history-focused episodes, the series' other main contributor was the progressive author and broadcaster Rhoda Power. Rhoda, with input from her sister, the historian Elaine Power, was responsible for the content of the history-focused episodes in the series. With titles such as, "Near-Man Becomes a Cave-Dweller" and an emphasis on long timespans and global perspectives, evolutionary themes also regularly featured in the history episodes written by Power.[3]

Beginning with *How Things Began*, and in particular the experimental approach to bringing radio broadcasts to life pioneered by Rhoda Power, this chapter outlines the post-war growth of educational broadcasts on evolutionary science. As with the above example, the chapter highlights how, given the didactic aims of educational broadcasting and the perceived malleability of children's minds, anti-evolution groups targeted schools broadcasts more than any other type of BBC content on evolution. Following a brief history of schools and adult educational broadcasting, the chapter then introduces the launch of televised schools broadcasts in 1957, and adult educational broadcasts in 1963; exploring how the creation of programmes on evolutionary theory, which were aimed at specific target audiences affected the format and narrative style of early pedagogical efforts.

The chapter highlights the diversity of approaches adopted by media producers across both TV and radio in early educational content on evolution, and contrasts them against the narrative approach favoured in mainstream productions for a general audience as introduced in the previous chapters. The chapter finishes by considering the launch of the BBC partnership with the newly founded distance-learning institute, the Open University (OU), in 1971. Early content on evolution made as part of this BBC-OU partnership illustrated theoretical aspects of the subject using the latest case studies from academia, often featuring the scientists involved in the original research.

[2] "What is a Fossil", "Why Fossils are Found on Land" and "How Living Things Change", written by Honor Wyatt, September–October 1944, BBC Home Service. Honor Wyatt was well connected in literary circles, particularly through her friendship with the writers Robert Graves and Laura Riding; notably moving in the same social circles as Jacob Bronowski among many others. William Graves, "News", *Gravesiana: The Journal of the Robert Graves Society* (1998), 2, 247–249, http://www.robertgraves.org/issues/15/6367_article_247.pdf *(last accessed 16/06/2021)*.

[3] "Near-Man Becomes a Cave-Dweller," *How Things Began*, written by Rhoda Power, first broadcast January 19, 1943, BBC Home Service.

Religious Contestation of Early Educational Radio on Evolution

How Things Began was typical of school broadcasts in the 1940s, which under the direction of Mary Somerville (Chap. 2), had grown greatly in both quantity and ambition since humble beginnings in the 1920s.[4] The old austere radio lecture, which proved unpopular with students and teachers alike, had been reinvented, and the "illustrated talk" had become the modus operandi for BBC schools broadcasts (Fig. 4.1).[5] The writers of *How Things Began*, in particular Rhoda Power, who had joined the BBC School Broadcasting Department permanently in 1939, were at the heart of these changes.[6] Encouraged by Somerville and Richard Palmer, a former lecturer in education and lead for schools science broadcasting, Power pioneered the introduction of features, such as sound effects, period music, narrative dialogue and dramatised re-enactment of historical events.[7] In doing so, Power effectively invented the illustrated talk format.[8] *How Things Began* was one of the first long-running series to use this approach, with most episodes set-up as conversations between a girl, her brother and their uncle. The children—George and Alice—played the role

[4] Having taken over as head of schools radio broadcasts in 1929, the BBC promoted Somerville to assistant controller of the whole talks division in 1947.

[5] Research commissioned by the BBC, and funded by the Carnegie UK Trust in 1927, had found that education officials, teachers and pupils all found schools broadcasts lacking in various regards. Carnegie United Kingdom Trustees. *Educational Broadcasting: Report of a Special Investigation in the County of Kent.* Carnegie UK Trustees, 1928. See David Crook, 'School Broadcasting in the United Kingdom: An Exploratory History', *Journal of Educational Administration and History* 39:3 (2007): 217–26, https://doi.org/10.1080/00220620701698341.

[6] Interested readers can watch Rhoda Power and her illustrated talks approach in action in the studio in the 1943 British Council promotional film *Lessons From the Air* available online at: http://film-directory.britishcouncil.org/lessons-from-the-air (British Council Film Collection, *last accessed 20/06/2021*).

[7] David Crook, 'School Broadcasting in the United Kingdom: An Exploratory History', *Journal of Educational Administration and History* 39:3 (2007): 218–219, https://doi.org/10.1080/00220620701698341; and Anne Pimlott Baker, "Power, Rhoda Dolores De Poer", *Oxford Dictionary of National Biography* (2006) https://doi.org/10.1093/ref:odnb/58995.

[8] Kenneth Fawdry, *Everything But Alf Garnett: A Personal View of BBC School Broadcasting* (British Broadcasting Corporation, 1974), 45. For more on Rhoda Power's career and lasting influence at the BBC see Laura Carter, 'Rhoda Power, BBC Radio, and Mass Education, 1922–1957', *Revue Française de Civilisation Britannique. French Journal of British Studies*, no. XXVI–I (2021), https://doi.org/10.4000/rfcb.7316.

Fig. 4.1 Producers from the schools department rehearse an early dramatised script, left to right: Brian Michie, Mary Somerville, Rhoda Power and George Dixon. It was not until the mid-1940s that Mary Somerville had the budget to hire professional actors. Richard Palmer, *School Broadcasting in Britain*. BBC, 1947

of "enquirers" asking questions about natural and human history, providing a more appealing narrative arc and framework in which the didactic segments of the programme could be situated. Scientifically speaking, the evolutionary content in *How Things Began* was not particularly cutting-edge, as Richard Palmer later reflected, "the evolutionary story was entirely new ground for the majority of schools listening".[9] The series, targeted at 10–14-year-olds, was light on specific evolutionary concepts and

[9] Richard Palmer, *School Broadcasting In-Britain* (British Broadcasting Corporation, 1947), 110; R. D. Bramwell, 'SCHOOL BROADCASTING AND CURRICULUM INNOVATION IN ENGLISH SCHOOLS', *Journal of Educational Administration and History* 10:1 (1978): 41–45, https://doi.org/10.1080/0022062780100106.

mechanisms, favouring the more engaging grand narrative of the evolutionary epic over more didactic modes of teaching biology.[10]

As we saw in Chap. 3, the BBC was familiar with co-ordinated campaigns by anti-evolutionist groups.[11] Thus in their responses, BBC officials often used stock text and gave complainants short shrift. For example, in response to another complaint about *How Things Began*, the BBC answered that content was "based on the best scientific advice available" and that "it is thought that broadcasting is an unsuitable medium for scientific controversy".[12] While complaints against adult broadcasts with evolutionary themes often attempted to contest specific scientific content, and demand airtime for opposing fringe or pseudo-scientific viewpoints, complaints against school broadcasts on evolution usually took a different approach.[13] Although often being written by the same anti-evolution groups, when dealing with school broadcasts, complaints overwhelmingly focused on the perceived moral implications of teaching children evolutionary science. Several letters from the period focused on the malleability and innocence of children's impressionable minds, with one complainant highlighting the confusion that will be caused by science broadcasts that contradict Sunday school lessons.[14] Others felt that evolutionary theory itself overstepped the boundaries of scientific fact, arguing that a theological outline of man's ancestry and creation was more appropriate for a

[10] Bernard Lightman, 'Evolution for Young Victorians', *Science & Education* 21:7 (2012): 1015–34, https://doi.org/10.1007/s11191-010-9333-0; Melanie Keene, *Science in Wonderland: The Scientific Fairy Tales of Victorian Britain* (Oxford University Press, 2015).

[11] Highlighting that the 1944 complaints against *How Things Began* were part of a planned and co-ordinated campaign is the fact that the letter from the Bible Testimony Fellowship was dated September 15, 1944, over a week prior to the first of the episodes on evolution actually being aired.

[12] Letter from G. J. Willoughby (Director, BBC Secretariat) to A. E. Hughes (Chairman, The Bible League), December 4, 1944. BBC-WA R41/51/1. All BBC copyright content reproduced courtesy of the British Broadcasting Corporation. All rights reserved.

[13] For example, in December 1942 the BBC Director of Talks received a letter from Evolution Protest Movement founders L. A. Davies and D. Dewar, which attempted to leverage on their expertise as "scientific workers" to, in point-by-point form with citations, contest each of the misleading scientific statements put forward in the series *Man's Place in Nature*. Letter from L. A. Davies and D. Dewar to BBC Director of Talks, December 15, 1942, BBC-WA R41/51/1.

[14] Letter from Chas Payne to BBC Director of Religious Broadcasting, December 2, 1946, BBC-WA R41/51/2. Harold Munro Fox's broadcasts also received similar complaints about the perceived damage done to vulnerable children's minds. See Max Long, "Secrets of Nature: Mass media and public understandings of the natural world in interwar Britain" (University of Cambridge, PhD Thesis).

young audience. As the Association of Catholic Trade Unionists put it, the BBC's approach was "contrary to all accepted Christian principles of juvenile education".[15]

On occasion, religious figures did contest schools content on evolution on philosophical rather than theological grounds. For example, following the earlier schools series *Biology in the Service of Man* (1937), the Jesuit scholar Cyril C. Martindale wrote a lengthy letter, which while praising Professor Harold Munro Fox's talks critiqued some of the normative assumptions and teleological language he used.[16] After stressing that "Catholics are committed to no dogma as to the pre-history of what is now the human body", Martindale complained about zoologist Munro Fox's imprecise use of language and attempt to make biological statements on what were for him philosophical questions. In particular, Martindale called out Munro Fox's directional and biased account of cleverness, stating that the phrase "'became cleverer' contains a positive suggestion that the mind of the man was only an improved edition of the animal mind".[17] Although echoing criticisms of scientism levelled at popular radio content (Chap. 3), highbrow intellectually nuanced complaints against educational biology programmes were rare and vastly outnumbered by letters from those with an a priori anti-evolutionist position.

Rather than retreating completely from controversial subjects, the experimental and progressive Schools Department repeatedly maintained that children had a great capacity to deal with grey areas and uncertainty. To take one of several examples, in the autumn of 1949 in a regular slot on religion and philosophy aimed at pupils aged 16–18, the Schools Department ran a series of three broadcasts exploring the philosophical implications and impacts of science on religion.[18] Despite being followed by an episode of *Religion and Philosophy* in which the Bishop of Bristol

[15] Letter from F. G. Foy (Honorary Secretary, Association of Catholic Trade Unionists) to BBC Director of Schools Broadcasting, October 26, 1949. BBC-WA R41/51/2.

[16] In addition to opposition from anti-evolution groups, anti-vaccination groups also opposed Harold Munro Fox's broadcasts on ideological grounds. Max Long, "Secrets of Nature".

[17] Letter from C. C. Martindale (Campion Hall, Oxford University) to BBC 'Science for Schools Talks', August 3, 1939. BBC-WA R51/523/2. Underlined for emphasis in the original.

[18] The three episodes were "The Scientific Spirit," *Religion and Philosophy (For the Schools)*, presented by K. G. Collier, September 26, 1949, BBC Home Service; "The New Scepticism: The Nineteenth Century," *Religion and Philosophy (For the Schools)*, presented by K. G. Collier, October 10, 1949, BBC Home Service; and "The New Scepticism: The

gave the argument for a belief in God in the age of science, the original broadcasts still managed to elicit angry letters from a small minority of religious viewers.[19] As with many other areas of moral panic and conflict in society, children—portrayed as innocent, malleable vessels—were at the centre of most of the co-ordinated campaigns against BBC content, with broadcasts on evolution no exception.[20] The potential influence of each school broadcast on evolution was heightened by the fact that these recordings were often repeated year on year. For example, the main content from some episodes of *How Things Began* was repeated annually from the early 1940s until 1967.[21]

In 1947, Richard Palmer published a book called *School Broadcasting in Britain*, which for each subject outlined the progress made, the rationale for the approach taken and the utility of educational radio in the classroom setting. Televised educational broadcasts were on the horizon when Palmer published his book. However, before we turn to TV, we must know a little more about the history of educational broadcasting more generally at the BBC.

A Short History of BBC Educational Broadcasting

From its earliest incarnation education was a central part of the BBC, inextricably associated with the "Reithian values" of the organisation.[22] During the 1920s, educational broadcasting was one of only four programme-

Twentieth Century," *Religion and Philosophy (For the Schools)*, presented by K. G. Collier, October 17, 1949, BBC Home Service.

[19] For an example of how the BBC responded to such letters, see: Letter from Mary P. Usher (Secretariat, Religious Broadcasting) to Mrs C. Stubbs, October 29, 1949. BBC-WA R41/51/2.

[20] For more on some of the most high-profile moral campaigns against broadcasting content in the post-war decades, and the disproportionate focus on children and young people see: Mary Whitehouse and Jonny Trunk, *Ban This Filth!: Letters from the Mary Whitehouse Archive* (Faber & Faber, 2012).

[21] The dramatised sequences from an episode of *How Things Began*, "Brain, eye and hand" written and presented by Honor Wyatt, first aired in November 1941, were still being used for school broadcasts into the early 1960s. For more on the longevity of scripts by Rhoda Power, see Carter, 'Rhoda Power, BBC Radio, and Mass Education, 1922–1957', 10.

[22] Following the Central Religious Advisory Committee, education was the second area to be overseen by an external advisory body, when the Central Educational Advisory Committee was appointed in August 1924. For more on Reith and early educational efforts at the BBC, see Asa Briggs, *The History of Broadcasting in the United Kingdom: Volume II: The Golden Age of Wireless* (Oxford University Press, 1965), Part II, Chap. 4.

producing departments alongside Talks, Music and Productions.[23] Despite the buy-in from Reith and initial excitement at the potential of radio for educational broadcasting, the novelty of the wireless, combined with the conservatism of many school teachers meant that initial progress with schools broadcasting was slow. However, by incorporating educationalists, both as staff and external advisers, and by conducting research in the classroom setting and then refining formats and content in response, the popularity of the "For the Schools" segment on the BBC National Programme rose during the 1930s.[24] Driven largely by Mary Somerville, who had been appointed Director of School Broadcasting in 1929, Schools' growing popularity was rewarded with larger budgets and the expansion of dedicated sections. School broadcasts came to dominate the output of the Education Department as they broadened their scope with enduring segments.[25] In 1935 the BBC estimated that some 6% of pamphlets produced to accompany schools broadcasts were being sent to adult listeners in private homes.[26]

In spite of severe limitations on budgets and facilities, the war years saw both adult and schools education continue to grow in popularity. In 1942 it was estimated that just over 50% of schools in the UK were utilising the service, and by 1944 the BBC Schools Department had grown to 30

[23] Early educational efforts were closely connected to the Christian religious values at the centre of Reith's vision for the BBC. Indeed until 1933 when it became a dedicated department, religious broadcasting was overseen by the BBC's Director of Education, J. C. Stobbart. Michael Bailey, '"He Who Has Ears to Hear, Let Him Hear": Christian Pedagogy and Religious Broadcasting During the Inter-War Period', *Westminster Papers in Communication and Culture* 4:1 (2017), 5, https://doi.org/10.16997/wpcc.70.

[24] This track was set by the first Director of Education, J. C. Stobart who was seconded from the Board of Education; while the first classroom research undertaken by the BBC begun in 1927 and was funded by the Carnegie Trust. Mary Somerville, "How School Broadcasting Grew Up," in Palmer, *School Broadcasting In-Britain*. For more on the development of schools science broadcasting during the war, and how Somerville and Palmer saw its relation to the development of a general policy on science at the BBC see: Talks, Science, Files 2A & 2B, R51/523/3 & 4, BBC-WA.

[25] Reflecting this dominance the Adult Education section was moved from the Education Department to Talks in 1928.

[26] Crook, 'School Broadcasting in the United Kingdom', 218–222; Briggs, *The History of Broadcasting in the United Kingdom*; and Allan Jones, 'Mary Adams and the Producer's Role in Early BBC Science Broadcasts', *Public Understanding of Science* 21:8 (2012): 968–83, https://doi.org/10.1177/0963662511419450..

staff.[27] Science broadcasts in particular did well in the post-war years, as many state-run schools were under-resourced and poorly equipped for science lessons. As Richard Palmer reflected in 1947, "[i]n short in many schools in England and Wales, the broadcasts began to determine, to a very large extent, what science was taught".[28]

The close oversight of several expert committees helped ensure that when Mary Somerville retired as Director of School Broadcasting in 1947, the direction of travel remained much the same under her successor Richmond Postgate, previously of the BBC Forces Education Unit.[29] The first of three central themes of the post-war science syllabus focused on applied science. Aligned with the pre-war science and society approach, it was an attempt to avoid presenting science as a "water-tight compartment" linking it with other relevant subjects.[30] *How Things Began* was the prime example of such a "cross-bred series", beginning firmly in the domain of science with the evolution of life on earth, and as the series progressed, charting the development of human civilisations transitioning to history, via archaeology and anthropology.[31] The second central theme of the science syllabus was human biology, an area often neglected by embarrassed schoolteachers, which included "the exciting evolutionary story of reproduction in the backboned animals".[32] Here, again, *How Things Began* was the flagship broadcast, with the evolutionary episodes planned by a former university teacher of zoology.[33] The final theme of the science syllabus was the "methods and attitudes of the scientist", which

[27] Palmer, *School Broadcasting In-Britain*, 7; and David Crook, 'School Broadcasting in the United Kingdom: An Exploratory History', *Journal of Educational Administration and History* 39:3 (2007): 221, https://doi.org/10.1080/00220620701698341.

[28] Palmer, 99.

[29] As well as the School Broadcasting Council and its subgroups such as the Science Sub-Committee, other committees, such as the Central Religious Advisory Committee regularly reviewed shows. For more on how Christian pedagogy shaped BBC content see: Bailey, '"He Who Has Ears to Hear, Let Him Hear"'. Although Somerville officially retired, she remained with the BBC becoming Assistant Controller of the Talks Department in April 1947. BBC, *Annual Report and Accounts 1946–47* (HMSO, 1947), 13; "News." *Western Daily Press and Bristol Mirror* [Yeovil, England] April 11, 1947: 3; "Richmond Postgate." *Times* [London, England] December 26, 1991: 18; Palmer, *School Broadcasting In-Britain*, 99; and Crook, 'School Broadcasting in the United Kingdom' 221.

[30] Palmer, *School Broadcasting In-Britain*, 104.

[31] Palmer, 108.

[32] Palmer, 101.

[33] Palmer, 109.

aimed to show how scientific knowledge and methods could be used in everyday life.[34] This theme was brought to life in *How Things Began* by the consistent insistence of the children Alice and George, "intended as representatives of the audience" for more evidence of the scientific facts relayed to them by the adults.[35]

In addition to *Talks to Sixth Forms*, aimed at 16–18-year-olds studying for their A-Levels,[36] educational science broadcasts for schools were split between two streams, *General Science* (1940–1961), and *Science and the Community* (1939–1964). Despite the title, *Science and the Community* was predominantly about technology, and most often in the decade following the war focused on industrial processes of mass production, vital to reconstruction efforts. Even with this focus, on occasion evolutionary themes and concepts found their way into broadcasts, for example in the episode "The Story of our Farm Animals" broadcast as part of the 1952 spring-term series, *Science and the Larder*. *General Science*, on the other hand, featured numerous segments on biology and evolution.

Students listening to either series at a registered school received a pamphlet to accompany the broadcasts each term.[37] The pamphlets helped to alleviate some of the limitations of radio by providing relevant images, such as microscope slides of cells (Fig. 4.2), relevant graphs and charts, and photographs of scientists and other historical figures narrated by actors on the broadcast recordings.[38] So, when in January 1951 children in isolated rural schools across Britain heard a French-accented Louis Pasteur talking about his important microbial discoveries, a photograph of

[34] Palmer, 100.

[35] Palmer, 108–109.

[36] The General Certificate of Education (GCE) Advanced Level, or A-Levels, is subject-specific academic qualification introduced in 1951 to replace the Higher School Certificate. Undertaken between the ages 16 and 18, they are aimed at those with academic aptitude, who perform well at the General Certificate of Education (GCE) Ordinary Level, or O-Level exams taken at 16. While O-Levels were replaced by the General Certificate of Secondary Education (GCSE) in 1988, A-Levels remain the primary exam undertaken by school-leavers in the UK wishing to go on to university.

[37] A complete series of all BBC Schools pamphlets for both radio and television for the years 1926–1979, can be found in The BBC Broadcasts for Schools Collection, held by the Institute for Education at University College London. For more information see: https://libguides.ioe.ac.uk/bbcsch *(last accessed 18/02/20)*.

[38] When pamphlets had been discontinued during the war, the difficulties of successfully teaching the biology course "when the children were deprived of the essential diagrams and illustrations" had been noted. *BBC Handbook 1940* (BBC, 1940), 70.

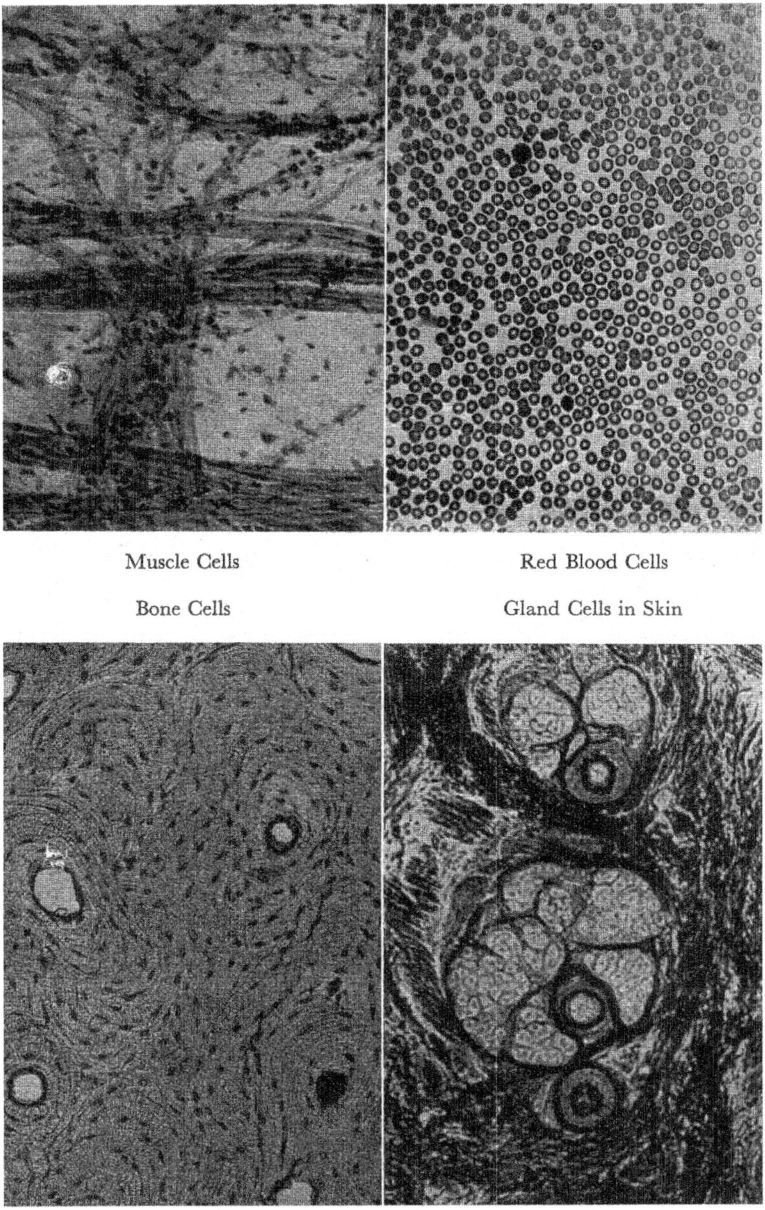

Fig. 4.2 Microscope images of human body cells, created by Flatters and Garnett Ltd., featured in the pamphlet for "Growing Up" episode broadcast, February 14, 1951. *General Science, BBC Broadcasts to Schools*, Spring Term 1951 (BBC, 1951), 21

Table 4.1 General Science broadcast schedule, spring term 1951. Broadcast on the BBC Home Service on Wednesdays between 11:00 and 11:18 a.m. General Science, BBC Broadcasts to Schools, spring term 1951 (BBC, 1951), 1

UNIT I	
The world of microbes	
17 January	Leeuwenhoek discovers microbes
24 January	Microbes in the air
31 January	Microbes in water
7 February	Life in a pond
UNIT II	
Growing up	
14 February	What are you made of?
21 February	How water animals begin to develop
28 February	How land animals begin to develop
7 March	Boys, girls and twins
14 March	Healthy growth

Pasteur in his laboratory aided their imaginations.[39] The pamphlets also included simple experiments that the pupils could do, either in the classroom or at home. Following the Pasteur broadcast in the *World of Microbes* series (Table 4.1), students were encouraged to experiment fermenting apple peel and record the results—"[c]an you smell the alcohol?" Throughout, the pragmatic always anchored the theoretical, so following these experiments children were encouraged to reflect on why canning meat and vegetables lengthens their shelf life.[40]

Aligning with the school term, content in the *General Science* series was usually split into two distinct units from the same discipline (Table 4.1). Given its centrality in the syllabus, biology featured more heavily than chemistry or physics. However, evolution as a distinct explanatory framework did not feature. Given the target audience were 10–14-year-olds, instead, biological processes were taught in a more systematic and pragmatic way, rather than by being explicitly situated within a larger explanatory framework. Thus, while in the summer term of 1951 pupils learnt about animal physiology and behaviour, progressing from simpler to more complex organisms, the evolution of species or phylogenetic trees was not introduced. On the other hand, *How Things Began*, which was aimed at

[39] *General Science, BBC Broadcasts to Schools*, Spring Term 1951 (BBC, 1951), 8.

[40] *General Science*, 11. Encouraging pupils to do practical experiments had been popular since the inter-war period, see Max Long, "Secrets of Nature."

the same age-group, was more narrative-driven, chronologically structured and introduced some basic general mechanisms, such as natural selection. Thus for many children growing up in the UK in the post-war years, *How Things Began* would have been the first time they encountered the "story" of evolution.

Content for pupils aged 16–18, which had a tradition of delivering talks by eminent academics, dealt with evolution in more detail. For example, in 1935–1936, Professor of Genetics at the University of London, J. B. S. Haldane had delivered a wide-ranging series of lectures, including one on "Plants and Animals in Human History", which discussed how *Homo sapiens* have affected the evolution of other species.[41] By the post-war years, series of talks were more structured and thematically grouped than Haldane's wide-ranging pre-war efforts. A concentrated focus on evolution for sixth-formers was first broadcast in 1956, via a series of six talks by Professor of Zoology at University College London, George P. Wells (1901–1985). Wells was the son of the science fiction pioneer H. G. Wells, and in 1929–1930 he had co-authored the pioneering three-volume popular book *The Science of Life* with his father and Julian Huxley.[42] The series of talks delivered by Wells covered similar content to Huxley's lectures aimed at adults in the same period, utilising a chronological progressive narrative, and finishing with reflections on man's future evolution.[43] However, again, as with the younger cohort listening to *How*

[41] *Talks for Sixth Forms: Science*, "Plants and Animals in Human History", written and presented by J. B. S. Haldane, BBC National Programme, February 7, 1936. Haldane continued to give occasional lectures for sixth-form audiences until December 1938, when he spoke in the *Forceful Thinkers* series on Charles Darwin. His original lectures were repeated in 1943, and during the war, he regularly featured as an expert panellist on *The Brains Trust*.

[42] The three-volume, nine book, *The Science of Life* was one of the first comprehensive biology textbooks ever published. Serialised in 31 fortnightly instalments the book was a great success and contained, in Book Four—"The How and Why of Development and Evolution," one of the earliest outlines of Huxley's version of the evolutionary synthesis. For more on the writing of *The Science of Life* and in particular how it diverted Huxley's attention away from his scientific research and academic career, see Steindór J. Erlingsson, 'The Costs of Being a Restless Intellect: Julian Huxley's Popular and Scientific Career in the 1920s', *Studies in History and Philosophy of Science Part C: Studies in History and Philosophy of Biological and Biomedical Sciences* 40:2 (2009): 101–8, https://doi.org/10.1016/j.shpsc.2009.03.006.

[43] The six episodes broadcast in January and February 1956 were titled: "The Changing Earth", "The Story of Evolution", "How Can We Explain Evolution?", "The Genes in

Things Began, it was not just on such purely biologically focused series that older pupils were introduced to evolution.

Sixth formers were also introduced to the more teleological aspects of the evolutionary epic in the 1953 series of talks, *The Idea of Progress*. In particular, in the third episode on "Science and Civilisation", botanist Eric Ashby (1904–1992) spoke about scientists' role in driving social progress, and in episode seven palaeontologist, Frederick Zeuner (1905–1963) spoke on "Progress and Evolution". Across this wide-ranging series, a focus on humanity's propensity for "progress"—our cultural evolution—was frequently situated within a teleological telling of the evolutionary epic.[44] As with the 1943 series *Reshaping Man's Heritage* (Chap. 2), *The Idea of Progress* was driven by a post-war desire to reaffirm societal belief in the progress of humanity, and to situate science at the centre of this vision for our species' future.[45]

Thus, for both secondary school and sixth-form pupils, the dry and didactic evolutionary biology content of post-war educational science broadcasts was reinforced by evolutionary themes in other series. Often covering large timescales and integrating expert voices from many academic disciplines, these non-science series centred on teleological narratives that placed humans, and in particular Western civilisation, at the pinnacle of progressive forces. While doing so they often loosely referenced evolution and utilised its associated terminology, conflating biological evolutionary theory with the political philosophy of progressivism. Although not explicitly labelled scientific humanism, this approach presented science as the driving force of social progress, and utilised the

Heredity", "The Genes of Evolution" and "Evolution in the Present and the Future", all written and presented by George P. Wells, BBC Home Service, 1956.

[44] This series, which is worthy of much more in-depth analysis by future scholars than I can afford it here, also featured in the sixth episode an exploration on the theme of disillusionment with progress in the works of H. G. Wells, and a final episode providing reflections on the series, which featured the politician Enoch Powell, and the physician and popular writer Alex Comfort (Chap. 5).

[45] Although episodes in the series reflected on the concept of progress across a wide range of subjects, from economics to political democracy, repeatedly episodes called for the application of "rational" or "scientific" processes across these domains, and the airtime afforded more critical perspectives was limited. While such positive perspectives of progress were popular in the period, there were lots of more critical perspectives from other public intellectuals, for examples see: Morris Ginsberg, *The Idea of Progress: A Revaluation* (Beacon Press, 1953); and Bertrand Russell, "The Idea of Progress: Ethics and Biology", *The Manchester Guardian*, March 14, 1953, 4.

familiar narrative appeal of the evolutionary epic. Even the in-depth coverage of science content on evolution, while lacking the sweeping grand narrative found in popular broadcasts, was still situated within a techno-progressive narrative. Palmer's three themes of the post-war radio science syllabus—applied science, biology and the scientific method—clearly aligned with Huxley's synthetic vision for humanities' evolutionary future.

Educating the Grown-Ups

Although radio was considered to have great potential for adult education, it was not until the popularity of school broadcasts among adults was realised that the BBC gave it concerted attention.[46] The Adult Education Section was separated from school broadcasting in 1927, and in 1930 merged with the general Talks Department. While some of the more esoteric output of Talks was educational in some regard, Adult Education remained a distinct section within the department, developing a specific approach, which distinguished their content from more general lectures or talks in the period.[47] During the inter-war years, the focus for Adult Education was on creating a more "educated democracy".[48] Shaped by experts in adult education, via the Central Council for Broadcast Adult Education, the approach was paternalistic, sometimes patronising and overwhelmingly centred on attempts to raise the intellect of the working classes.[49] In the early 1930s, regional adult education councils were established, and following an experiment by the Kent education authority,

[46] Crook, 'School Broadcasting in the United Kingdom'.

[47] Briggs, *The History of Broadcasting in the United Kingdom: Volume II: The Golden Age of Wireless*, 142–7.

[48] See the 1919 *Report on Adult Education*, as introduced in Briggs, 186; and *New Ventures in Broadcasting: A Study in Adult Education* (BBC, 1928). The latter includes the findings of a joint committee tasked with exploring the possibilities of broadcasting as a means of general adult education, an overview of which can be found in the *BBC Handbook 1929* (BBC, 1929). For wider context on debates around the BBC's role in a modern democracy see: D. L. LeMahieu, *A Culture for Democracy: Mass Communication and the Cultivated Mind in Britain Between the Wars* (Clarendon Press, 1988).

[49] In 1936 the Council was renamed the Central Committee for Group Listening. *BBC Handbook 1938* (BBC, 1938), 23. For more on the paternalistic approach of the BBC towards education in the period, see Jones, 'Mary Adams and the Producer's Role in Early BBC Science Broadcasts', 970–71.

regular listening groups grew in popularity nationwide.[50] Often taking place in public libraries, village halls or hosted by other civil organisations, such as the Y.M.C.A., groups met regularly to listen to and discuss educational series of interest to their members. A Group Leader who directed discussion, with support from a local Education Officer, accompanying pamphlets and the BBC's own educational magazine *The Listener*, would lead sessions.[51]

Prior to the Second World War, Mary Adams produced most Adult education science content. Following the success of her series of talks, *Problems of Heredity* (1928), Adams joined the adult education section of the BBC in 1930.[52] Although she produced a wide range of science content, evolutionary and eugenic themes featured in many of the adult education series Adams was behind in the early 1930s. Notably, one of the first series she was involved with, *A1 or C3? The Future of the Race* (1930), was an exploration of current eugenic ideas and thoughts. With provocative titles such as, "Shall the Unfit Survive?" the talks explored the purported biological underpinnings of race science and, as was common during the period, connected them to pressing social and political issues of the day. Adams' view that universal access to education was an essential component of a successful eugenic agenda dovetailed neatly into the paternalistic and class-driven focus that the Adult Education section had already established.[53]

Adams left radio talks in 1936 for the new Television Department. Her departure coincided with a dip in adult education output, as the section

[50] Following the success of the schools broadcasting experiments in 1927, and again funded by the Carnegie UK Trustees, Kent Education Authority bought a number of radio sets, which they loaned to trial listening groups. In 1928–1929, there were 160 registered listening groups, and by 1937, the BBC was regularly selling in the region of 10,000–60,000 pamphlets to accompany adult education series. *BBC Yearbook 1930* (BBC, 1930), 86–87; and *BBC Annual 1937* (BBC, 1937), 85.

[51] *BBC Handbook 1939* (BBC, 1939), 77-80. For more on the structure of educational talks series, a general overview of scientific content, and some examples from the pamphlets that accompanied them see: Jones, 'Mary Adams and the Producer's Role in Early BBC Science Broadcasts' 971–74. For an in-depth study of Listening Groups in the early 1930s see: Alexandra Lawrie, 'The BBC, Group Listening, and "The Changing World"', *Media History* 25:3 (2019): 279–91, https://doi.org/10.1080/13688804.2019.1623664.

[52] The talks were published as a book the following year: Mary Adams, *Six Talks on Heredity* (Heffer and Sons Ltd, 1929)

[53] Jones, 'Mary Adams and the Producer's Role in Early BBC Science Broadcasts' 974.

struggled to compete for funds as the UK shifted to a wartime footing.[54] Following the war, as radio ownership became more widespread (Fig. 3.1), listening groups declined in popularity.[55] The establishment of the Forces Education Unit (FEU) in 1945 gave adult education a much needed renewed focus and impetus.[56] The FEU was tasked with shifting the morale boosting educational broadcasts made for the forces during wartime, to the peacetime aim of "fitting the men and women of our fighting services for the transition to civilian life".[57] Echoing the post-war schools syllabus, FEU split their science content into two streams: pure and applied. By the late 1940s, influenced by Archibald Clow's functionalist approach to science, applied content made up the overwhelming majority of FEU science broadcasts. While FEU geared some of this applied content towards practical knowledge that discharged personnel may utilise in future careers,[58] most focused on communicating the positive and integral role that applied science played in society.[59] In actuality, the gap between the pure and applied streams was not always evident. Although the limited evolutionary content on FEU broadcasts was "pure science", such episodes often appeared in supposedly applied series. For example, a lecture by the zoologist George P. Wells on "Growth and Heredity", a pre-cursor to his 1956 series introduced earlier in this chapter, was one of six talks he gave in the applied series *Science and Everyday Life* in late 1948.[60]

[54] Allan Jones, 'Science in the Making: 1930s Citizen Science on the BBC', *History of Education*, https://doi.org/10.1080/0046760X.2020.1712620: 15–16.

[55] The Area Councils and associated BBC funding for group listening was discontinued in 1947. *BBC Annual Report and Accounts for the Year 1946–47* (BBC, 1948), 32; and Lawrie, 'The BBC, Group Listening, and "The Changing World"', 288.

[56] Sometimes referred to as the Forces Educational Broadcasts Unit, or the Services Educational Unit

[57] *The Listener*, March 22, 1945. In 1945 forces educational broadcasts accounted for approximately 5.5% of the output of the BBC Light Programme, but by 1948 this had dropped to just 1.6%, broadcasts specifically for the forces continued until 1952. Asa Briggs, *The History of Broadcasting in the United Kingdom: Volume IV: Sound and Vision* (OUP, 1995), 57; 736–38.

[58] "Forces Educational Broadcast Science and Everyday Life", *Radio Times*, 1304, October 8, 1948, 15.

[59] Jared Robert Keller, 'A Scientific Impresario: Archie Clow, Science Communication and BBC Radio, 1945–1970' (Imperial College London, PhD Thesis, 2017), http://spiral.imperial.ac.uk/handle/10044/1/57504.

[60] The full series of talks by George P. Wells was "The Living Units of the Body", "The Circulation of the Blood", "The Living Engine", "Food and Digestion", "The kidneys and

The lack of a comprehensive syllabus for forces education meant that biology content was fragmentary, and when evolution did appear, it was limited to isolated episodes. Furthermore, unlike the schools schedule, adult education broadcasts in the decade following the war lacked series with overarching progressive narratives, like *How Things Began* and *The Idea of Progress*. Yet while adult education broadcasts in the period may not have used the longer teleological narrative of the evolutionary epic, they did have a dominant normative framing, which centred a contemporary lens on scientific materialism and progressivism. This approach to science communication was not the Whiggish preserve of scientists alone, as exemplified by Archibald Clow's FEU series *Science and History* (1945–1946), which regularly featured other experts like museum curators and historians.[61] Despite the broader range of experts in FEU shows, adult education science content in the post-war period continued to promote Huxley's vision of scientific humanism.

Despite the successes of both schools and adult educational radio, and the re-launch of BBC television in June 1946, limited post-war funds meant that televised educational broadcasts were not prioritised in the decade following the war. In the early 1950s, with the spectre of the first commercial television stations on the horizon, the BBC began to experiment with televised educational science content.[62] However, despite this belated investment they were not the first broadcaster to regularly air televised science for schools.

what they do", and "Growth and Heredity", *Science and Everyday Life,* Forces Educational Broadcasts, BBC Light Programme, November to December 1948.

[61] *Science and History* explored scientific developments from the last 200 years and was innovative because Clow paired each scientific talk with a corresponding *History and Science* talk, usually delivered by a historian, broadcaster or other relevant expert. Keller, 'A Scientific Impresario', 68–69.

[62] In 1949, Lord Beveridge began a large official enquiry into the future landscape of broadcasting in Britain. Published in January 1951 the lengthy report made over 100 recommendations, and began a conversation about monopoly in broadcasting. Following a subsequent Conservative Government White Paper of May 1952, and protracted political debate the Television Act 1954 was passed, which for the first time allowed regional commercial franchises to broadcast under the regulation of the Independent Television Authority. For a detailed account of Beveridge, and the subsequent committees which ultimately led to commercial broadcasting in the UK, see Briggs, *The History of Broadcasting in the United Kingdom*, 1995, in particular, 340–360 and 803–909.

Televised Educational Output

In May 1957, the first regional franchise of the newly established commercial Independent Television (ITV) network, London-based Associated-Rediffusion (AR) usurped BBC plans and began televised broadcasts for schools.[63] Although ITV was a commercial network created to break the BBC's monopoly, the Television Act (1954) which led to its creation mandated a public service remit for the new regional franchises.[64] This hybrid approach to commercial broadcasting meant that regional ITV franchises had to strike a balance between market forces and public service.[65] Under guidance from their own Schools Committee, AR moved fast to ensure they were the first to broadcast content for schools. During the first term of experimental programmes, AR broadcast one show per day aimed at students 14 and above, to an estimated total of just 125 schools.[66] AR dedicated Thursday's slot to science, with leading scientists appearing on a *Year of Observation* (1957) to discuss the latest findings in relation to the International Geophysical Year then underway. AR made a strategic decision to target Secondary Modern Schools, who received lower funding per student, as their middle- and working-class students were deemed less intellectual and not expected to go on to university.[67]

Following a long experimental phase and pressure from the School Broadcasting Council, in the autumn of 1957 the BBC launched their own schools television service, initially broadcasting two to three shows per week.[68] Despite local authorities equipping around 300 schools with

[63] "School Doubts on Television," *The Times*, December 20, 1956, 5. Unfortunately, when Associated-Rediffusion lost its franchise in 1968 most of its programme archives and other associated records were destroyed. The British Film Institutes National Archive holds most still in existence.

[64] HM Government, *The Television Act*, Bill 127 (HMSO, 1954) and HM Government, *Lords Amendments to the Television Bill*, Bill 146 (HMSO, 1954), 3. For discussion of the public service remit see: Bernard Sendall, *Independent Television in Britain: Origin and Foundation 1946–62* (Springer, 1982), 31–36.

[65] Catherine Johnson and Rob Turnock, *ITV Cultures: Independent Television Over Fifty Years: Independent Television Over Fifty Years* (Open University Press, 2005), 196–98.

[66] Although AR only broadcast in the London region, the shows were also simultaneously broadcast on the Associated Television (ATV) franchise in the Midlands region. Sendall, *Independent Television in Britain*, 275.

[67] Sendall, *Independent Television in Britain*, 272–277.

[68] British Broadcasting Corporation, *BBC Annual Report and Accounts 1956–57* (HMSO, 1957), 44–45, 139; British Broadcasting Corporation, *BBC Handbook 1957* (BBC, 1957), 90; Sendall; Carter, 'Rhoda Power, BBC Radio, and Mass Education, 1922–1957', 9–10.

televisions, access to a suitable set was an issue for most.[69] By the end of 1957, only 1100 schools were registered to receive the new BBC television schools service, which was in stark contrast to the 29,000 schools, 73% of the total, that were regularly listening to BBC schools radio series.[70] Given this disparity in reach and the greater resources required to make television, the newly formed BBC Schools television unit had an experimental brief to work with schools, educators and local authorities to develop a curriculum best suited to the medium. In particular, science was deemed an area "in which schools badly need[ed] help". Thus, the series *Science and Life* was one of two main shows in the trial, and a natural history show aimed at 11–12-year-olds followed in the first term of 1958.[71] Initially evolution did not receive any concerted attention, and was dealt with piecemeal across the *Science and Life* and *Natural History* segments, with mini-series such as "Living in Water", "How Your Body Works" and "How Animals Move" broadcast during 1958–1959.[72]

When launching televised science for schools, both the BBC and AR highlighted the great potential of the visual element of television to aid in the classroom. As the secretary of the School Broadcasting Council, Richard Steele put it; "television's power to open a window for children on the living, real world around them" presented a great opportunity to aid the "resources of the teacher and his laboratory".[73] This acute awareness of the visual potential of television for science in the classroom echoed concurrent discussions about the broadcasting of science for general audiences. When, in 1949, the Royal Society and the British Association for the Advancement of Science (BAAS) had petitioned for closer liaison and a dedicated BBC science department (Chap. 3), senior producers defended

[69] R. C. Steele, "BBC Television Service for Schools," *Radio Times*, September 20, 1957, 3.

[70] British Broadcasting Corporation, *BBC Handbook 1957* (BBC, 1957), 89 and 109.

[71] R. C. Steele, "BBC Television Service for Schools," *Radio Times*, September 20, 1957, 3; and British Broadcasting Corporation, *BBC Annual Report and Accounts 1958–59* (HMSO, 1959), 143.

[72] British Broadcasting Corporation, *BBC Annual Report and Accounts 1958–59* (HMSO, 1959), 143

[73] R. C. Steele, "BBC Television Service for Schools," *Radio Times*, September 20, 1957, 3. Steele's article mirrored what the Chairman of AR had told their annual general meeting in December 1957, when he first announced that they would be launching schools broadcasts in May 1957. "Television for Schools," *The Times*, December 28, 1956, 7; and Sendall, *Independent Television in Britain*, 269.

their autonomy by outlining the tacit knowledge required for successfully communicating science via broadcast media.[74] In detailing how the translational processes involved in successfully communicating science differed by mediums, BBC television producers James McCloy and Aubrey Singer both emphasised that the visual aspect of television was central to the selection of scientific subject matter.[75]

In September 1959, ITV launched its flagship schools science programme, *Discovery* (1959–1967).[76] Produced by Granada Television, the franchise holder for the north of England, the series of 30-minute lectures, experiments and demonstrations aimed to bring famous scientists into classrooms.[77] Unlike AR, Granada's early schools series was aimed at sixth-form students, aged 16–18, "intellectually speaking ... the *élite*". As the head of educational broadcasting at Granada, former journalist Sir Gerald Barry explained, these children were the ones that would become future leaders across all fields.[78] Developed in partnership with the BAAS, with editorial input from the Editor of *New Scientist*, and incorporating input from head teachers, the series featured stand-alone talks by scientists, initially independent of any syllabus.[79]

[74] Timothy Boon, *Films of Fact: A History of Science in Documentary Films and Television* (Wallflower Press, 2008), 221–224.

[75] James McCloy to Cecil McGivern (Controller of Programmes, Television), "Presentation of Science by Television", October 7, 1958; and "Science Broadcasts", Meeting Notes, October 22, 1958, 1, T16/263, BBC-WA.

[76] For a wealth of information and resources on *Discovery* and other schools broadcasting from the period, see the amateur fan website "Broadcasts for Schools", compiled by Ben Clarke, http://www.broadcastforschools.co.uk/site/Discovery_(TV) *(last accessed 17/03/21)*.

[77] The interested reader can get an idea of the format and approach taken by watching a short clip online, taken from the lecture "Spermatozoa" written and presented by Lord Rothschild, originally broadcast June 2, 1960, on Granada/ITV: https://www.youtube.com/watch?v=qm_tZlUsgsw *(last accessed 17/03/2021)*.

[78] Gerald Barry, "A Word About Discovery," *Discovery: Developments in Science* (Methuen & Co., 1961), preface. Highlighting just how close elite public intellectual networks were in the period, in 1965 Barry had co-edited a book with Jacob Bronowski and Julian Huxley. Gerald Barry, Jacob Bronowski, James Fisher and Julian Huxley, *Health and Wealth Man's Fight Against Sickness and Want* (Macdonald, 1965).

[79] "Meeting Minutes" *Granada T.V. Network Ltd.*, April 6, 1959, 1131, School's Programmes (ITV Archive, Leeds); "Science Programme for Sixth-Forms" *Granada Memo*, June 26, 1959, 1131, School's Programmes (ITV Archive, Leeds); and "New General Brief," *Discovery*, August 4, 1960, 1135, School's Programmes (ITV Archive, Leeds).

During planning, evolution was one of six subject areas identified as having good potential, and thus featured in four of the initial ten episodes.[80] In week three, Cambridge University Professor of Organic Chemistry, Sir Alexander Todd (1907–1997)[81] introduced "Nucleic Acids", while the following week, Sidney Brenner (1927–2019) also of Cambridge built on Todd's talk, speaking about DNA and the emerging understanding of genetic molecules' role in inheritance. In lecture seven, a third Cantabrigian, zoologist, Sir James Gray (1891–1975) introduced the statistical improbability of life in relation to the Second Law of Thermodynamics:

> [W]hen we look at the development of an egg, such as that of a newt, the situation is not quite so clear. The whole process seems much more like the development of organized structure from a relatively simpler system. … Now the plan of the embryo is determined by the long evolutionary history of the newt, but it is carried out by a machine developed within the egg itself. … It is not at all easy to see how this sort of thing can be usefully described in terms of a system which is constantly trying to reach a state of maximum physical probability. There seems to be some directive principle at work.[82]

Gray's lecture was frank about the limitations of current biological knowledge, and unlike contemporary mainstream series on evolution, such as *Five Hundred Million Years* (Chap. 3) went into detail about current debates in the field. In particular, Gray critiqued natural selections' ability to act as the sole explanatory mechanism for evolution and concluded that his own feeling was that "some indication of directive evolution may eventually be found".[83] Thus, while Gray dedicated a large portion of the lecture to the problems of imposing directional narratives on evolution, as we see in the above quote, he not only occasionally used teleological language himself, but also shared his own belief in a directional objective for evolution.

[80] The six programme subjects suggested in the initial planning meeting were "Space – Matter & Motion", "Chemistry of Fibres", "Large Molecules", "Relativity", "Atomic Structure & Atomic Energy" and "Evolution". "Meeting Minutes" *Granada T.V. Network Ltd.*, April 6, 1959, 1131, School's Programmes (ITV Archive, Leeds).

[81] For more on Todd's other BBC broadcasts, including his reflections on the challenges of representing chemistry via a visual medium, see Soraya de Chadarevian, *Designs for Life: Molecular Biology After World War II* (Cambridge University Press, 2002), 151–53.

[82] James Gray, "The Science of Life", *Discovery: Developments in Science* (Methuen & Co., 1961), 12–13. The lecture, originally broadcast on November 12, 1959, was one of 15 from the first four series of *Discovery* published verbatim in a popular book format in 1961.

[83] James Gray, "The Science of Life", 16.

Although Gray largely employed directive and design-oriented language metaphorically, he situated it in the normative frame of the evolutionary epic.[84] Gray explicitly placed biological evolution as a successive phase of cosmic evolution and attempted to grasp at the "Essence of Life":

> To some extent a living organism can be described in terms of molecules and units of energy but—like a beautiful picture—it is something more. To discover just what this something is, provides the spur which drives biology along.[85]

Gray infused his version of the evolutionary epic with the touchstones of Huxley's scientific humanism; emphasising our common ancestry with other animals, while expounding our eminence, and humbly discussing the limits of our knowledge while promoting the belief that science will continue to elevate us further.

Reinforcing the themes introduced in Gray's lecture, in week nine of the series Professor of Animal Genetics at Edinburgh University, Conrad Waddington (1905–1975) tackled "the appearance of design in living things". Waddington began by sketching an outline of man's long history of observing design in animal physiology, introducing Jean-Baptiste Lamarck's theory on the inheritance of acquired characteristics, and describing how Darwin's mechanism of natural selection usurped it in the late nineteenth century. Waddington then introduced the unit of inheritance that was unknown to Darwin, the gene using his own research on the fruit fly (*Drosophila melanogaster*). He then outlined his controversial theory of genetic assimilation, whereby a physical change by a species in response to their environment, over many generations becomes encoded in their DNA.[86] Like Gray, Waddington was open about the limits of cur-

[84] Gray's use of teleological language largely as a narrative device employed to move the "plot" of his lecture forward reflects the findings of more recent work exploring the frequency of such language in natural history programming. See Meryl Aldridge and Robert Dingwall, 'Teleology on Television?: Implicit Models of Evolution in Broadcast Wildlife and Nature Programmes', *European Journal of Communication* 18:4 (2003): 435–53, https://doi.org/10.1177/0267323103184001.

[85] James Gray, "The Science of Life", 17.

[86] Although controversial at the time, Waddington's work continues to illicit interest, with genetic assimilation—albeit via a revised conception—being argued for as a useful concept that can help to broaden the scope of contemporary genetic models of phenotypic evolution. Massimo Pigliucci, Courtney J. Murren and Carl D. Schlichting, 'Phenotypic Plasticity and Evolution by Genetic Assimilation', *Journal of Experimental Biology* 209:12 (2006): 2362–67. For more on Waddington's research and how it related to his anti-reductionist

rent scientific understanding, and in contrast to the more reductive content on evolution for general audiences, discussed competing ideas in the neo-Darwinian landscape. Waddington's lecture highlighted the tension when trying to communicate complex evolutionary concepts between using precise, neutral scientific language and more agentive teleological wording. Even though the lecture was largely a scientific deconstruction of the appearance of design in animals, the limitations of the format and the desire to present an arresting narrative meant that teleological—"perfectly designed for digging", and agentive—"natural selection picks out those which happen to be appropriate"—language abounded.[87]

After these initial successes, evolution remained a central feature of *Discovery*. Unlike the BBC approach, whereby educators and science communicators mediated content on evolution, the first series of *Discovery* gave students direct access to cutting-edge debates in professional biology. The lectures neatly encapsulated the pressing concerns of top British biologists at the time, as they grappled with the explosion in genetics research and its implications for mechanisms of inheritance.[88]

The first comprehensive treatment of evolution on schools television broadcasting was *The Evolution of Living Things*, part of the BBC *Science and Life* series that aired in the spring term of 1960. Aimed at 12–14-year-olds, Professor of Zoology, William Bullough (1914–2010) a regular fixture on radio school broadcasts on natural history and evolution since the mid-1950s, wrote and presented the series.[89] After episodes introducing the concept of change over time and the fossil record, *The Evolution of Living*

beliefs see: James F. Stark, 'Anti-Reductionism at the Confluence of Philosophy and Science: Arthur Koestler and the Biological Periphery', *Notes and Records: The Royal Society Journal of the History of Science* 70:3 (2016): 279–80, https://doi.org/10.1098/rsnr.2016.0021.

[87] Conrad Waddington, "Evolution: the appearance of design in living things", *Discovery: Developments in Science* (Methuen & Co., 1961), 19 and 21.

[88] See Richard G. Delisle ed., *The Darwinian Tradition in Context: Research Programs in Evolutionary Biology* (Springer International Publishing, 2017), in particular section two "Crossing the Boundaries: Between Non-Darwinian and Darwinian".

[89] William S. Bullough was Professor of Zoology at Birkbeck College, University of London from 1953 until 1982. Gaining his reputation with research on the reproductive cycle of vertebrates, in particular Starlings (*Sturnus vulgaris*) and later working on the mitogenic function of sex hormones in mammals, in addition to his broadcast work, he also published several popular textbooks throughout his career. See "Sorby Fellowship: Dr. W. S. Bullough," *Nature* 157, 187 (1946). https://doi.org/10.1038/157187d0; "Obituary: William S Bullough," Birkbeck University of London website, http://www.bbk.ac.uk/about-us/obituaries/obituary-william-s-bullough *(last accessed 11/03/2021)*; W. S. Bullough "The Starling (Sturnus Vulgaris L.) and Foot-and-Mouth Disease."

Things sequentially dealt with the emergence of each major vertebrate class in the animal kingdom.[90] Finally, the closing two episodes covered extinction caused by humans and artificial selection, and mechanisms of inheritance.

The structure of the series matched closely the approach taken by the most popular biology textbooks of the period.[91] The pedagogical aims and the amount of airtime afforded to the subject resulted in the series avoiding some of the grander teleological arguments found in broadcasts for general audiences. Yet this standard classroom treatment of evolution, which dealt with organisms chronologically and with increasing complexity, was inherently directional. This is not to make a value judgement on the occasional use of teleological language evident in both *Discovery* and *The Evolution of Living Things*, where despite the extended airtime, the input of leading scientists and an open acknowledgement of the pitfalls of such language, the use of purposeful, design-oriented and directive language abounded. The continued use of directional language across these educational broadcasts on evolution alerts us to how embedded such language is in our modes and ways of narrating a story, particularly one which aims to impart such complex information.[92]

Televised output for schools and in particular science content, on both ITV and the BBC, proved popular. In 1961, the BBC Television School Broadcasting Department moved into its own studio, which included a dedicated laboratory for science broadcasts.[93] Other regional franchises in

Proceedings of the Royal Society of London. Series B, 131:862 (1942): 1–12; and William S. Bullough, *The Evolution of Differentiation* (Academic Press Inc., 1967).

[90] The individual episodes were "The Evolution of Living Things", "Records in the Rocks", "The First Fishes", "The Colonisation of the Land", "The Age of Reptiles", "The Death of the Reptiles", "The Age of Mammals", "The Age of Man", "Evolution Today" and "How Animals Change". Each episode was initially aired twice and the whole series was repeated again in the autumn term of 1961.

[91] The schedule of this series, and the biology radio broadcasts introduced earlier in the chapter, reflect the structure used in biology textbooks and classroom syllabi in the period. See, for example, Nuffield Foundation, *Nuffield Biology Text I: Introducing Living Things* and *Nuffield Biology Text II: Life and Living Processes* (Longmans/Penguin Books, 1966).

[92] Michael Ruse, 'Evolutionary Biology and the Question of Teleology', *Studies in History and Philosophy of Science Part C: Studies in History and Philosophy of Biological and Biomedical Sciences*, Special Issue: Replaying the Tape of Life: Evolution and Historical Explanation, 58 (2016), 100, https://doi.org/10.1016/j.shpsc.2015.12.001.

[93] In the school year 1961–1962, of 35 televised series broadcast nationally for schools (primary, secondary and sixth form), 13 were science based, and of these 6 had biology content. British Broadcasting Corporation, *BBC Annual Report and Accounts 1961–62* (HMSO, 1962), 33, 145.

Table 4.2 Episodes of series 22 of *Discovery* aired in the autumn of 1966. "Evolution: Discovery 22" Independent Television programmes for schools, autumn term 1966 (Granada TV Network, 1966)

	Programme	*Speaker*	*Aired*
1.	Adaptation	Prof A. J. Cain	Sept. 20 & 21
2.	Evolution does occur—The Fossil Record	Dr D. Nichols	Sept. 27 & 28
3.	Darwin's and Wallace's Theory	Caroline Medawar	Oct. 4 & 5
4.	Heredity	Prof J. Maynard Smith	Oct. 11 & 12
5.	Natural Selection in Action I: Animals	Prof P. M. Sheppard FRS	Oct. 18 & 19
6.	Natural Selection in Action II: Plants	Dr A. D. Bradshaw	Oct. 25 & 26
	Half-Term Repeat of Programme Six		Nov. 1 & 2
7.	The Effects of Natural Selection on Behaviour	Dr J. H. Crook	Nov. 8 & 9
8.	Speciation	Prof A. J. Cain	Nov. 15 & 16
9.	Evolution in Primates	Dr J. R. Napier	Nov. 22 & 23
10.	Evolution in Primates (cont.) Postscript on Evolution and the Future	Sir Peter Medawar FRS	Nov. 29 & 30

the ITV network quickly picked up *Discovery* and by 1964 it was available in all 13 broadcast regions.[94] As of 1964, around 8500 schools were receiving televised broadcasts for schools, up from 6000 in 1963, and the Department of Education and Science highlighted the great potential of television for education in a government report.[95]

Having departed from its non-syllabus focused approach, Granada dedicated series 22 of *Discovery* in 1966 exclusively to evolution (Tab. 4.2). Although sticking to the formula of using an eminent academic for each

[94] "Discovery 16" *Independent Television Programmes for Schools, Autumn Term 1964* (Granada TV Network, 1964), front matter.

[95] The number of schools registered for televised schools output continued to grow steadily throughout the early 1960s. The BBC reported 5000 schools registered for television broadcasts in 1962–1963 and ITV reported an audience of 4132 schools in January 1963. "Education in 1964", *Report of The Department of Education and Science* (HMSO, 1965), 30, 118; British Broadcasting Corporation, *BBC Handbook 1964* (BBC, 1964), 60; and "Audience – ITV Schools Programmes," *Memo from Weltman/SRE to SLS*, January 31, 1963. 1245 – Education, Schools' Programmes (ITV Archive, Leeds).

episode, the first Professor of Genetics at the University of Liverpool, Philip Sheppard oversaw the series development.[96] Sheppard was a well-known contributor to popular science media in the period, regularly featuring on BBC radio and television.

The 1966 *Discovery* series on evolution, in both its structure and content, was a pared back and simplified version of a typical introductory university class on evolution.[97] In the introduction to the series, Sheppard laid out the central tenets of evolutionary theory and the most exciting recent discoveries. Beginning with Darwin, Wallace and to a lesser extent Lamarck and Mendel, and then outlining the subsequent twentieth-century explosion in mathematical and field approaches, Sheppard's account was a typical mid-century version of the evolutionary synthesis.[98] With most of the contributors loosely basing their TV lectures on popular accounts they had already published.[99] Sheppard's own popular textbook, *Natural Selection and Heredity* (1958), not only provided some of the content for his episode in the series, but also shaped the narrative arc of the series as a whole.[100]

In keeping with the episodes on evolution in series one of *Discovery* (1957) and the BBC's *The Evolution of Living Things* (1960), the 1966 *Discovery* series presented a distinctly British version of the evolutionary

[96] The series was produced by Granada TV's Peter Mullings who also worked on the later series of *Zoo Time* after Desmond Morris had left the show (see Chap. 3): 0174 Discovery (ITV Archive, Leeds). For an overview of Sheppard's career, in particular his genetics research on polymorphism in butterflies, moths and snails see: Arthur Cain, "Professor P. M. Sheppard, F.R.S.", *Heredity*, 40 (1978), 317–319. https://doi.org/10.1038/hdy.1978.33

[97] Mathematical and chemical content was kept to an absolute minimum, so that the series was still intelligible to those without specialist knowledge of these subjects. "Evolution: Discovery 22" *Independent Television Programmes for Schools, Autumn Term 1966* (Granada TV Network, 1966), 14.

[98] Ronald Fisher, *The genetical theory of natural selection* (Oxford University Press, 1930); and H. B. D. Kettlewell, "Further selection experiments on industrial melanism in the Lepidoptera", *Heredity*, 10, 287–301.

[99] Draft scripts for most of the series' episodes can be found in the ITV Archive (Leeds) in folders "0174 – Discovery" and "0178 – Discovery".

[100] The biology collection of the Hutchinson University Library series was co-edited by Arthur Cain and featured books by three of the ten scientists in this series of *Discovery*, namely: Arthur Cain, *Animal Species and their Evolution* (Hutchinson University Library, 1966); David Nichols, *Echinoderms* (Hutchinson University Library, 1962); and Philip Sheppard, *Natural Selection and Heredity* (Hutchinson University Library, 1958).

synthesis. While it avoided the more grandiose and utopian aspects as most frequently promoted by Julian Huxley, it gave sixth-form viewers a detailed insight into how biological research had developed beyond the original mechanisms proposed by Darwin and Wallace.[101] In incorporating contemporary genetics and other emergent areas of biological study, these sixth-form series amended the narrative arc of earlier, younger-age educational series. The evolutionary epic however remained central. *Discovery's* structure (Table 4.2), which bookended contemporary scientific content with episodes on the history of evolutionary ideas and the future of evolution and humanity, quickly became the default way to structure television series on the subject.[102]

The *Discovery* series on evolution was a great success, receiving bigger audiences than previous series of the programme, with 71% of those providing feedback rating the content as good or very good. The majority of the feedback from teachers centred on the visual aspects of the series, with many highlighting how visual aids, models and footage of animal species around the world were welcome additions to an abstract part of the syllabus that was difficult to illustrate in a classroom setting. Teachers deemed episode three on Darwin and Wallace the least effective because, unlike BBC Education who by this point were using dramatisation and historical re-enactments, the *Discovery* episode relied largely on still images.[103] ITV repeated the series in the autumn term of 1967, but despite the positive feedback, unlike earlier series of *Discovery*, no book version followed, and in 1968 *Discovery* quietly disappeared from the schedule.

[101] For more on utopianism in the modern evolutionary synthesis see: Maurizio Esposito, 'Utopianism in the British Evolutionary Synthesis', *Studies in History and Philosophy of Science Part C: Studies in History and Philosophy of Biological and Biomedical Sciences*, Defining Darwinism: One Hundred and Fifty Years of Debate, 42:1 (2011): 40–49, https://doi.org/10.1016/j.shpsc.2010.11.007.

[102] While recent pedagogical work has demonstrated that the use of historical narratives can be useful in teaching science, particularly the history of biology, scholars argue to be successful such an approach must focus on the evolution of fundamental concepts, not the life of one or two famous scientists. Daniel Gamito-Marques, 'In Praise of a Historical Storytelling Approach in Science Education', *Isis* 111:3 (2020): 582–87, https://doi.org/10.1086/711126.

[103] Granada Television, "Discovery – Autumn Term 1966, Evolution Series", 0174, Discovery (ITV Archive, Leeds).

Evolution on Adult Education Television

Although lagging behind by a few years, once adult educational broadcasting received more dedicated attention from producers, televised science followed a similar trajectory to schools. Following the recommendations of the Pilkington Committee on broadcasting (1960) and the subsequent Television (Higher Education Service) Bill (1961), the BBC launched a dedicated adult education television slot in October 1963.[104] An estimated 7,000,000 viewers tuned in to at least one series of the first BBC1 adult education offerings during 1963–1964.[105] The BBC had experimented with adult education science broadcasts since May 1961 when they launched *Science on Saturday*; a series consisting of repeats of science programmes and syndicated content produced by the US National Academy of Sciences. Curated by subject, the BBC aimed *Science on Saturday* (1961–1962) at "those who wish[ed] to use television for the planned study of science by adults".[106]

In its early adult education TV the BBC were again spurred on by competition from ITV franchises as they expanded into adult education during the early 1960s.[107] Initial BBC TV content drew on the expertise of the radio adult education department, in particular the Forces Education Unit. Both the BBC and ITV franchises relied heavily on the input of advisory committees and direct input from university departments of adult education.[108] While recapturing some of the experimental spirit of early science education radio broadcasts, BBC adult education television attempted to avoid the condescending tone of earlier radio content, and developed an andragogic approach, distinct from schools content.[109]

The amount of adult educational content produced by the BBC quickly expanded, with the launch in April 1964 of the BBC's second channel,

[104] Harry Pilkington, *Report of the Committee on Broadcasting, 1960*, Cmnd 1753 (HMSO, 1962); and *Television (Higher Education Service) Act*, Bill 87 (HMSO, 1961). For more on the Pilkington Committee and its findings see: Asa Briggs, *The History of Broadcasting in the United Kingdom: Volume V: Competition* (OUP, 1995), 257–303.

[105] British Broadcasting Corporation, *BBC Annual Report and Accounts 1963–64* (HMSO, 1962), 12. All BBC copyright content reproduced courtesy of the British Broadcasting Corporation. All rights reserved.

[106] British Broadcasting Corporation, *BBC Annual Report and Accounts 1961–62* (HMSO, 1962), 35

[107] "Education in 1964", *Report of The Department of Education and Science* (HMSO, 1965), 76.

[108] "Education in 1964", 118.

[109] Jones, 'Science in the Making', 2.

BBC2, its first foray into colour television. Created as a home for more highbrow, culturally relevant content in-line with the findings of the Pilkington Committee, BBC2 was a natural home for both adult education and more general popular science series. The first adult education TV content on biology was the programme *The Science of Man* (1963–1966) created by BBC science producer, James McCloy in 1963. Airing in early 1965, the fourth series was on "Heredity and Evolution", and although focused more narrowly on human biology, it included five of the ten experts that featured on the ITV Schools' *Discovery* series on evolution just one year later. The overlap of expertise between these two series, produced by rival broadcasters, reflects the coalescence during the period of an elite class of scientists, who regularly undertook work on the popularisation of science.

One of the main concerns of the Pilkington Committee with regard to Adult Education had been the lack of college and university-level broadcast content in the UK.[110] Although the BBC had been attempting to address this shortfall since 1963, it was not until the late 1960s and the realisation of long-mooted plans to establish a distance-learning university that they adopted a more comprehensive approach.[111]

The BBC and the Open University[112]

The BBC launched a partnership with the newly founded distance-learning institute, the Open University (OU) in January 1971.[113] Initially, the aim

[110] This is in stark contrast to the USA, where universities had long been directly involved with televised science initiatives; from the 1940s, medical schools had used closed-circuit systems to broadcast surgical procedures to students, educational broadcasts such as *The Johns Hopkins Science Review* (1948–1955) were created by higher education institutions, and albeit only making up a small percentage of schedules, by 1953 many major US cities had one educational station. Marcel Chotkowski LaFollette, *Science on American Television: A History* (University of Chicago Press, 2013), 12–15, 25–26, 59.

[111] The first vision for a "Wireless University" had been drawn up in 1926 by the first head of BBC Education, J. C. Stobart, who's ambitious "Wellsian" plan envisioned two-year courses on specific subjects taught either wholly or partially by radio. Briggs, *The History of Broadcasting in the United Kingdom: Volume II: The Golden Age of Wireless*, 188.

[112] Some of the text in this section was adapted, with permission from Alexander Hall, 'Evolution on the Small Screen: Reflections on Media, Science, and Religion in Twentieth-Century Britain', in *Identity in a Secular Age: Science, Religion, and Public Perceptions*, ed. Fern Elsdon-Baker and Bernard Lightman (University of Pittsburgh Press, 2020).

[113] A "university of the air" was proposed by Harold Wilson's Labour government in 1963, a key part of the Government's plans to modernise Britain, the project was driven by Minister of State for Education, Jennie Lee. The OU was established in 1969, and accepted its first students in January 1971. Pete Dorey, '"Well, Harold Insists on Having It!"—The Political

was to create radio and television content aimed at supplementing the written materials produced for OU students studying remotely across the UK. In terms of both the technical engineering requirements and its pedagogical aims, the partnership was hugely ambitious.[114] Science courses were a central component of the wide range of academic subjects offered by the new university, and with distance learners unable to access laboratories, there was an extra reliance on television to support the home experiment kits sent out to all students.[115]

Broadcast content on evolution to support both general science courses and higher-level biology modules was a feature from the outset, and over the coming decades, the BBC/OU partnership became the main producer of content on evolution in the UK.[116] The first BBC/OU broadcast, *Evolution by Natural Selection*, aired in June 1971 as part of the study materials for "Science: a foundation course", and was repeated regularly on the BBC until 1978.[117] The broadcasts produced for the science foundation course illustrated theoretical aspects using the latest academic case studies, often featuring the scientists involved in the original research. For example, in *Evolution by Natural Selection*, presented by OU scientists Dr Richard Holmes, Dr Peggy Varley and Professor Michael Pentz, the famous peppered moth experiment commonly used to illustrate the

Struggle to Establish The Open University, 1965–67', *Contemporary British History* 29 (2015), https://doi.org/10.1080/13619462.2014.981160. For a multimedia institutional history of the OU, see "Exhibition: The OU Story", *Open University Digital Archive*, https://www.open.ac.uk/library/digital-archive/exhibition/53 *(last accessed 18/03/21)*.

114 Anastasios Christodoulou, *Letter to the Director of Public Affairs*, January 15, 1971, R78/656/1, BBC-WA.

115 In this clip the first Dean of Science, Professor Michael Pentz (1924-1995) reflects on some of the early challenges, in particular the last-minute development of home experiment kits. "Inventing ways of teaching", *Open Forum 41: The First Ten Years*, 1979. Open University Digital Archive, https://www.open.ac.uk/library/digital-archive/clip/clip:sci_clipA *(last accessed 18/03/21)*.

116 Of the corpus of nearly 800 broadcasts on evolution introduced in Chap. 1 (Fig. 1.1), the BBC/OU partnership was responsible for over 100 individual programmes between 1971 and 2009. In addition to the televised content discussed here from 1971 onwards the BBC/OU partnership also produced educational radio, broadcast on BBC Radio 4.

117 "Module S100 – Science: a foundation course", *Open University*, 1971–1978. Open University Digital Archive, http://www.open.ac.uk/library/digital-archive/module/xcri:S100/study *(last accessed 18/03/21)*.

selective adaptation of a mutant gene was demonstrated using the video recording of the original experiment.[118]

Scientific experiments had been closely associated with film since its earliest days, both as an experimental aide and later reaching mass audiences in cinemas.[119] In addition to cinema shorts, since 1934 Gaumont-British Instructional had been making educational natural history films for classrooms of all ages. Informally advised by Julian Huxley, most of the films focused on laboratory experiments and up-close, microscopic or time-lapse footage of flora and fauna.[120] These theatrical and educational science films by pioneering British filmmakers developed and honed many of the technologies and filming techniques that the BBC adopted and used for televised science broadcasts.[121] As well as collaborating with these science-film specialists, BBC TV engineers and producers also pioneered their own approaches adapted to the specific limitations of the format.[122]

Influenced by pedagogical best practice, over and above emergent ideas of televisual best practice, the format for early BBC/OU science broadcasts was visually austere (Fig. 4.3) and the dialogue was didactic and factual without a clear narrative arc, a central feature of other TV science

[118] Interested readers can watch a clip of this part of the episode online at https://www.open.ac.uk/library/digital-archive/clip/clip:sci_clip14 *(last accessed 18/03/21)*. The mark-recapture experiments originally conducted by Dr Bernard Kettlewell during the 1950s quickly became a popular classroom example, and had previously featured in Professor Sheppard's October 1966 episode in the ITV *Discovery* series "Natural Selection in Action I: Animals" (Table 4.2). Following valid criticisms of the methodology the study was picked up by anti-evolutionist authors; however a subsequent comprehensive study has since vindicated Kettlewell's original findings, as originally presented in this OU television show. For more on the controversy see Stanley A. Rice, *Encyclopedia of Evolution* (Infobase Publishing, 2009), 306–309; and L.M. Cook et al., "Selective Bird Predation on the Peppered Moth: The Last Experiment of Michael Majerus," *Biology Letters* 8: 4 (2012): 609–12, https://doi.org/10.1098/rsbl.2011.1136.

[119] Boon, *Films of Fact*, 7–32.

[120] Max Long, 'The Ciné-Biologists: Natural History Film and the Co-Production of Knowledge in Interwar Britain', *The British Journal for the History of Science* 53:4 (2020): 527–51, https://doi.org/10.1017/S0007087420000370, 15.

[121] Gaumont-British Instructional were preceded in these endeavours by British Instructional Films (1919–1932), later filmmakers in this tradition who worked closely with the BBC, included the Macqueen Film Organisation (1950–1969) and Oxford Scientific Films (1968–present). Jean-Baptiste Gouyon, 'From Kearton to Attenborough: Fashioning the Telenaturalist's Identity', *History of Science* 49:1 (2011): 25–60, https://doi.org/10.1177/007327531104900102; and Long, 'The Ciné-Biologists'.

[122] Jean-Baptiste Gouyon, *BBC Wildlife Documentaries in the Age of Attenborough* (Springer Nature, 2019), 22, 185–193.

Fig. 4.3 A standard sparse and austere studio set-up for an Open University programme, circa early 1970s. Open University Digital Archive, http://www.open.ac.uk/library/digital-archive/image/image:000000010721, © The Open University

content of the period. As the scope and scale of televised education grew, particularly following the influx of funding via the BBC/OU partnership, we see in action the development of different sub-genres of popular science, as demarcations between fiction and non-fiction, pedagogic and popular approaches became more defined.[123] The result was that the broadcast study materials made under the BBC/OU partnership during the 1970s largely avoided explicitly progressive narratives and the more excessive use of teleological language found in both popular science content and some early school-level educational series.

From initially just two dedicated units within the OU General Science course, content on evolution and broadcasts to assist in its teaching

[123] For more on this demarcation and boundary work in popular science in print see: Felicity Mellor, "Between Fact and Fiction: Demarcating Science from Non-Science in Popular Physics Books," *Social Studies of Science* 33:4 (2003): 509–38, https://doi.org/10.2307/3182967.

continued to grow during the 1970s.[124] Most science modules in the period were supported by around 10 to 16 episodes, ranging from 10 to 30 minutes in length, which were repeated at least annually on BBC2. The OU launched a dedicated module on "Evolution" in 1992, accompanied by a complete ten-part documentary series.[125] Still written and narrated by OU experts, the content and narration remained close in approach to a classroom lecture.[126] However, by the 1990s, visually, the series were no longer disjointed lectures performed to camera, but looked more like mainstream documentaries, using natural history location shoots and computer graphics to bring technical concepts to life.

In addition to content made as part of science faculty courses, there was also a range of broadcasts that contained evolutionary themes made to support other OU courses. Most notably, the history of science module "Science and belief: from Copernicus to Darwin" (1974–1981) resulted in episodes covering how Charles Lyell's work and the voyage of the Beagle shaped Darwin's *On the Origin of Species*, and ideas caused by eighteenth-century fossil discoveries, particularly the works of Georges Cuvier, Jean-Baptiste Lamarck and William Buckland.[127] The BBC also made similar content on the history of evolution for the subsequent module, Science and belief: from Darwin to Einstein (1981–1987) (Chap. 7).

[124] Other modules in the 1970s which had accompanying broadcasts on evolution included, "Genetics" (S299, 1976–85), which had 16 accompanying BBC videos, covering subjects like "Mutations and Mutants" and "Genetic Engineering"; and "Ecology" (S323, 1974–85), from which the episode "Ants and acacias: an ecological study" can be watched online: https://www.open.ac.uk/library/digital-archive/program/video:00525_1130 *(last accessed 19/03/21)*. "Module List", Open University Digital Archive, https://www.open.ac.uk/library/digital-archive/module/list *(last accessed 19/03/21)*.

[125] All ten of the documentaries used for, "Evolution" (S365, 1992–2005), which were broadcast annually between 1992 and 2006, can be watched online: https://archive.org/details/OUBBCS365 *(last accessed 19/03/21)*.

[126] Writers and narrators on the video materials for the module "Evolution" (S365) included zoologist Caroline Pond, palaeo-biologist David Martill and evolutionary biologist Marion Petrie.

[127] "The Origin of The Origin of Species", listed in the *Radio Times* as simply "Origin of Species" was written and presented by the historian of biology Jonathan Hodge, while "The meaning of fossils", was written and presented by the historian of geology Martin Rudwick. "Module AMST283, The origin of the origin of species", https://www.open.ac.uk/library/digital-archive/program/video:00525_3119; and "Module AMST283, The meaning of fossils", https://www.open.ac.uk/library/digital-archive/program/video:00525_3115, Open University Digital Archive *(last accessed 19/03/21)*.

During the 1990s, the output of the BBC/OU partnership began expanding beyond course-related programmes, to making peak-time commissions for popular audiences.[128] The formats used for these popular BBC/OU documentaries were more akin to those used for regular BBC science broadcasts, and as such, utilised more storytelling devices such as historical re-enactments. With this came an emphasis on narrative arcs, larger synthetic frameworks and normative positions on the role and function of science in society. By 2009's BBC/OU production *Charles Darwin and the Tree of Life*, the reductive, mythic account of Darwin's life evident in other popular accounts on the history of evolution was front and centre (Chap. 8).[129]

This more recent style of BBC/OU commissions sits in stark contrast to the original co-produced educational materials, which were expansive, and through their specific didactic focus, limited in their forays into other disciplines. This strict demarcation by discipline meant that unlike more mainstream science broadcasts, grand synthetic narratives were limited. These programmes treated evolution as a bounded area of biological study, rather than the all-encompassing evolutionary narrative that had been promoted by popularisers such as Huxley since the inter-war period.

Conclusion

In this chapter we have seen how the expansive time afforded subjects, the disciplinary demarcation and the didactic emphasis of educational broadcasts on evolution resulted in radio and television shows that were less focused on narrative arcs and synthetic progressive visions than more

[128] This shift was influenced by several factors; the BBC was facing growing commercial pressures; academically informed science content continued to be popular with general audiences and triggered by growing student access to VCRs, in 1995, the BBC began "time-shifting", moving all OU educational content overnight on BBC2, which freed considerable prime time and daytime scheduling. "Small Screen Heroes: The OU and the BBC", *History of OU* Project, Open University Website, archived online at https://web.archive.org/web/20181201084844/http://www.open.ac.uk:80/researchprojects/historyofou/story/small-screen-heroes-the-ou-and-the-bbc *(last accessed 19/03/21)*; and Jean Seaton, *Pinkoes and Traitors: The BBC and the Nation, 1974–1987* (Profile Books, 2015), 304–328.

[129] *Charles Darwin and the Tree of Life*. Produced by Sacha Mirzoeff. Presented by David Attenborough. BBC, February 1, 2009.

popular iterations. Driven and overseen by educationalists, in the first decades of both schools and adult education, the BBC seldom taught content on biology within a wider evolutionary framework. Where evolution was taught as part of longer progressive narratives, such as in *How Things Began*, fundamentalist religious groups repeatedly raised worries about the implications of teaching evolutionary science to such malleable young minds. In later decades, reflecting wider societal trends with regard to the diminishing authority of religion in British society, contestation of educational content on evolution on religious grounds reduced.[130] A trend that is also reflected in the complete absence of religious figures and worldviews in educational science broadcasts, in contrast to earlier popular content such as *Five Hundred Million Years* (1958).

On the whole, schools and adult education broadcasts on evolution retained a largely corporate structure, closely aligned with classrooms, textbooks and university syllabi. However on occasion external influences, such as the competition brought by the launch of ITV Schools in 1957, allowed learners access to more individual, personal and sometimes fringe treatments of the subject. Even these limited instances, such as Waddington's lecture on the appearance of design in living things in the first series of *Discovery*, were largely gone from ITV Schools scheduling by the time it dedicated a whole series to evolution in 1966. The alignment of broadcasts with classroom syllabi meant that particularly for schools content, discussion of controversy, disagreement and wider issues were almost completely absent.

Given its attachment to university course content, the educational output produced under the BBC/OU partnership from 1971 onwards has unsurprisingly provided the most in-depth, scientific and up-to-date broadcasts on evolution. While made for a small-niche audience of students taking the specific OU module, this content regularly reached a wider daytime audience of the unemployed, housewives and other curious or bored TV owners. The lack of commercial pressures and the pedagogical aims of the BBC/OU partnership resulted in broadcasts centred not on narratives, but on imparting scientific knowledge and teaching experimental methods. However, as commercial pressures increased, and the BBC/OU partnership began making mainstream content, which utilised

[130] For detailed and comprehensive accounts of the many nuances of these societal trends, see Charles Taylor, *A Secular Age* (Harvard University Press, 2007); and Steve Bruce, *God Is Dead: Secularization in the West* (Blackwell, 2002).

common tropes and historical narratives about Darwin and evolution, BBC educational output on the subject increasingly adopted a more explicit progressive framing on evolutionary science.

On the surface educational broadcasts on evolution disrupt somewhat the top-level narrative arc of this book. In this area of broadcasting, the effect of synthetic thinkers like Julian Huxley was much more limited, with the importance of an overarching progressive evolutionary framework relegated by the priority of teaching biological knowledge. The continued oversight of educationalists meant that broadcast content rarely fell behind the latest academic thinking, and stereotypes and tropes were largely avoided. Finally, educational broadcasts did not deal with areas that caused controversy for specific interest groups, like human origins, or the future of evolution, especially at lower education levels.

However, when we reflect on the detail, we can see how the history of educational broadcasting on evolution can have a dual quality. The history presented in this chapter highlights (yet again) the plurality of approaches to communicating evolutionary themes via broadcast media. Yet we can also see that it was still part of broadcast media's wider role in reinforcing and normalising a progressive evolutionary framework within British society. Throughout the twentieth century, educational broadcasting remained an area where experimentation with new techniques and formats was encouraged. Techniques developed for educational broadcasts, such as Rhoda Power's historical re-enactments, repeatedly became normal devices used across many other genres and formats. When translated to other contexts, they often lost their critical focus and acted to reinforce stereotypical depictions of evolution (Chaps. 7 and 8).

Finally, while educational broadcasts on evolution were only explicitly progressive on rare occasions, they were always uncritical promoters of science. Even the limited number of BBC/OU broadcasts on the history and philosophy of evolution made since 1971 rarely discussed issues such as scientific materialism, progress in science, the scientific method or relativism. So while educational broadcasting may have not amplified conflict-centred narratives and evolutionary frameworks that encompassed the social and cosmic—at the same time it has been the single biggest promoter of science, the scientific method and the wonders of scientific progress within British broadcasting history.

References

Adams, Mary. *Six Talks on Heredity*, Heffer and Sons Ltd, 1929.
Aldridge, Meryl, and Robert Dingwall. 'Teleology on Television?: Implicit Models of Evolution in Broadcast Wildlife and Nature Programmes'. *European Journal of Communication* 18:4 (2003): 435–53. https://doi.org/10.1177/0267323103184001.
Bailey, Michael. '"He Who Has Ears to Hear, Let Him Hear": Christian Pedagogy and Religious Broadcasting During the Inter-War Period'. *Westminster Papers in Communication and Culture* 4:1 (2017). https://doi.org/10.16997/wpcc.70.
Baker, Anne Pimlott. "Power, Rhoda Dolores De Poer", *Oxford Dictionary of National Biography* (2006) https://doi.org/10.1093/ref:odnb/58995
Barry, Gerald. (Ed.) *Discovery: Developments in* Science, London: Methuen & Co., 1961. Barry, Gerald, Jacob Bronowski, James Fisher and Julian Huxley, *Health and Wealth Man's Fight Against Sickness and Want*, London: Macdonald, 1965.
BBC. *Annual Report and Accounts 1946–47*. HMSO, 1948.
———. Annual Report and Accounts 1961–62. HMSO, 1962.
———. Annual Report and Accounts 1963–64, HMSO. 1964.
———. BBC Annual 1937. BBC, 1937.
———. BBC Handbook 1929. BBC, 1929.
———. BBC Handbook 1938. BBC, 1938.
———. BBC Handbook 1939. BBC, 1939.
———. BBC Handbook 1940. BBC, 1940.
———. BBC Handbook 1964. BBC, 1964.
———. BBC Yearbook 1930. BBC, 1930.
———. General Science, BBC Broadcasts to Schools, Spring Term 1951. BBC, 1951.
———. New Ventures in Broadcasting: A Study in Adult Education. BBC, 1928.
Boon, Timothy. *Films of Fact: A History of Science in Documentary Films and Television*. Wallflower Press, 2008.
Bramwell, R. D. 'SCHOOL BROADCASTING AND CURRICULUM INNOVATION IN ENGLISH SCHOOLS'. *Journal of Educational Administration and History* 10:1 (1978): 41–45. https://doi.org/10.1080/0022062780100106.
Briggs, Asa. *The History of Broadcasting in the United Kingdom: Volume I: The Birth of Broadcasting*. Oxford University Press, 1961.
———. The History of Broadcasting in the United Kingdom: Volume II: The Golden Age of Wireless. Oxford University Press, 1965.
———. The History of Broadcasting in the United Kingdom: Volume IV: Sound and Vision. Oxford University Press, 1995.

———. The History of Broadcasting in the United Kingdom: Volume V: Competition. Oxford University Press, 1995.

Bruce, Steve. *God Is Dead: Secularization in the West.* Blackwell, 2002.

Bullough, William S. "The Starling (Sturnus Vulgaris L.) and Foot-and-Mouth Disease." *Proceedings of the Royal Society of London. Series B,* 131:862 (1942): 1–12.

———. *The Evolution of Differentiation.* Academic Press Inc., 1967.

Cain, Arthur. *Animal Species and their Evolution.* Hutchinson University Library, 1966.

———. "Professor P. M. Sheppard, F.R.S.", *Heredity,* 40 (1978), 317–319. https://doi.org/10.1038/hdy.1978.33.

Carnegie United Kingdom Trustees. *Educational Broadcasting: Report of a Special Investigation in the County of Kent.* Carnegie UK Trustees, 1928.

Carter, Laura. 'Rhoda Power, BBC Radio, and Mass Education, 1922–1957'. *Revue Française de Civilisation Britannique. French Journal of British Studies,* no. XXVI-I (2021). https://doi.org/10.4000/rfcb.7316.

Chadarevian, Soraya de. *Designs for Life: Molecular Biology After World War II.* Cambridge University Press, 2002.

Cook L.M. et al., "Selective Bird Predation on the Peppered Moth: The Last Experiment of Michael Majerus," *Biology Letters* 8:4 (2012): 609–12, https://doi.org/10.1098/rsbl.2011.1136.

Crook, David. 'School Broadcasting in the United Kingdom: An Exploratory History'. *Journal of Educational Administration and History* 39:3 (2007): 217–26. https://doi.org/10.1080/00220620701698341.

Delisle, Richard G. 'From Charles Darwin to the Evolutionary Synthesis: Weak and Diffused Connections Only'. In *The Darwinian Tradition in Context: Research Programs in Evolutionary Biology,* edited by Richard G. Delisle, 133–67. Cham: Springer International Publishing, 2017. https://doi.org/10.1007/978-3-319-69123-7_7.

"Discovery 16" Independent Television Programmes for Schools, Autumn Term 1964. Granada TV Network, 1964.

Dorey, Pete. '"Well, Harold Insists on Having It!"—The Political Struggle to Establish The Open University, 1965–67'. *Contemporary British History* 29 (2015). https://doi.org/10.1080/13619462.2014.981160.

Erlingsson, Steindór J. 'The Costs of Being a Restless Intellect: Julian Huxley's Popular and Scientific Career in the 1920s'. *Studies in History and Philosophy of Science Part C: Studies in History and Philosophy of Biological and Biomedical Sciences* 40:2 (2009): 101–8. https://doi.org/10.1016/j.shpsc.2009.03.006.

Esposito, Maurizio. 'Utopianism in the British Evolutionary Synthesis'. *Studies in History and Philosophy of Science Part C: Studies in History and Philosophy of Biological and Biomedical Sciences,* Defining Darwinism: One Hundred and Fifty Years of Debate, 42:1 (2011): 40–49. https://doi.org/10.1016/j.shpsc.2010.11.007.

"Evolution: Discovery 22", Independent Television Programmes for Schools, Autumn Term 1966. Granada TV Network, 1966.

Fawdry, Kenneth. *Everything But Alf Garnett: A Personal View of BBC School Broadcasting*. British Broadcasting Corporation, 1974.

Fisher, Ronald. *The genetical theory of natural* selection. Oxford University Press, 1930.

"Forces Educational Broadcast Science and Everyday Life", *Radio Times*, 1304, October 8, 1948, 15.

Gamito-Marques, Daniel. 'In Praise of a Historical Storytelling Approach in Science Education'. *Isis* 111:3 (2020): 582–87. https://doi.org/10.1086/711126.

Ginsberg, Morris. *The Idea of Progress: A Revaluation*, Boston: Beacon Press, 1953.

Graves, William. "News", *Gravesiana: The Journal of the Robert Graves Society* (1998), 2, 247–249, http://www.robertgraves.org/issues/15/6367_article_247.pdf

Gouyon, Jean-Baptiste. *BBC Wildlife Documentaries in the Age of Attenborough*. Springer Nature, 2019.

———. 'From Kearton to Attenborough: Fashioning the Telenaturalist's Identity'. *History of Science* 49:1 (2011): 25–60. https://doi.org/10.1177/007327531104900102.

Hall, Alexander. 'Evolution on the Small Screen: Reflections on Media, Science, and Religion in Twentieth-Century Britain'. In *Identity in a Secular Age: Science, Religion, and Public Perceptions*, edited by Fern Elsdon-Baker and Bernard Lightman. University of Pittsburgh Press, 2020.

HM Government, "Education in 1964", *Report of The Department of Education and Science*. HMSO, 1965.

———. Lords Amendments to the Television Bill, Bill 146. HMSO, 1954.

———. Television (Higher Education Service) Act, Bill 87. HMSO, 1961.

———. The Television Act, Bill 127. HMSO, 1954.

Johnson, Catherine and Turnock Rob. *Itv Cultures: Independent Television Over Fifty Years: Independent Television Over Fifty Years*. McGraw-Hill Education (UK), 2005.

Jones, Allan. 'Mary Adams and the Producer's Role in Early BBC Science Broadcasts'. *Public Understanding of Science* 21:8 (2012): 968–83. https://doi.org/10.1177/0963662511419450.

———. 'Science in the Making: 1930s Citizen Science on the BBC'. *History of Education* 49:3 (2020): 327–343. https://doi.org/10.1080/0046760X.2020.1712620.

Keene, Melanie. *Science in Wonderland: The Scientific Fairy Tales of Victorian Britain*, Oxford University Press, 2015.

Keller, Jared Robert. 'A Scientific Impresario: Archie Clow, Science Communication and BBC Radio, 1945–1970' (Imperial College London, PhD Thesis, 2017). http://spiral.imperial.ac.uk/handle/10044/1/57504.

Kettlewell, H. B. D. "Further selection experiments on industrial melanism in the Lepidoptera", *Heredity*, 10, 287–301.

LaFollette, Marcel Chotkowski. *Science on American Television: A History.* University of Chicago Press, 2013.

Lawrie, Alexandra. 'The BBC, Group Listening, and "The Changing World"'. *Media History* 25:3 (2019): 279–91. https://doi.org/10.1080/13688804.2019.1623664.

LeMahieu, D. L. *A Culture for Democracy: Mass Communication and the Cultivated Mind in Britain Between the Wars.* Clarendon Press, 1988.

Lightman, Bernard. 'Evolution for Young Victorians', *Science & Education* 21:7 (2012): 1015–34, https://doi.org/10.1007/s11191-010-9333-0.

Long, Max. 'The Ciné-Biologists: Natural History Film and the Co-Production of Knowledge in Interwar Britain'. *The British Journal for the History of Science* 53:4 (2020): 527–51. https://doi.org/10.1017/S0007087420000370.

———. "Secrets of Nature: Mass media and public understandings of the natural world in interwar Britain" (University of Cambridge, PhD Thesis).

Mellor, Felicity. "Between Fact and Fiction: Demarcating Science from Non-Science in Popular Physics Books," *Social Studies of Science* 33:4 (2003): 509–38, https://doi.org/10.2307/3182967.

"News." Western Daily Press and Bristol Mirror [Yeovil, England] April 11, 1947: 3.

Nichols, David. *Echinoderms.* Hutchinson University Library, 1962.

Nuffield Foundation, *Nuffield Biology Text I: Introducing Living Things* and *Nuffield Biology Text II: Life and Living Processes*, Longmans/Penguin Books, 1966.

"Obituary: William S Bullough," Birkbeck University of London website, https://www.bbk.ac.uk/about-us/obituaries/obituary-william-s-bullough *(last accessed 11/03/2021).*

Palmer, Richard. *School Broadcasting In-Britain.* British Broadcasting Corporation, 1947.

Pigliucci, Massimo, Courtney J. Murren, and Carl D. Schlichting. 'Phenotypic Plasticity and Evolution by Genetic Assimilation'. *Journal of Experimental Biology* 209:12 (2006): 2362–67. https://doi.org/10.1242/jeb.02070.

Pilkington, Harry. Report of the Committee on Broadcasting, 1960. Cmnd 1753, HMSO, 1962.

Rice, Stanley A. *Encyclopedia of Evolution*, Infobase Publishing, 2009.

"Richmond Postgate." Times [London, England] December 26, 1991: 18.

Ruse, Michael. 'Evolutionary Biology and the Question of Teleology'. *Studies in History and Philosophy of Science Part C: Studies in History and Philosophy of Biological and Biomedical Sciences*, Special Issue: Replaying the Tape of Life: Evolution and Historical Explanation, 58 (2016): 100–106. https://doi.org/10.1016/j.shpsc.2015.12.001.

Russell, Bertrand. "The Idea of Progress: Ethics and Biology", *The Manchester Guardian*, March 14, 1953, 4.

Seaton, Jean. *Pinkoes and Traitors: The BBC and the Nation, 1974–1987*. Profile Books, 2015.

Sendall, Bernard. *Independent Television in Britain: Origin and Foundation 1946–62*. Springer, 1982.

Sheppard, Philip. *Natural Selection and Heredity*. Hutchinson University Library, 1958.

"Sorby Fellowship: Dr. W. S. Bullough," *Nature* 157, 187 (1946). https://doi.org/10.1038/157187d0.

Steele, R. C. "BBC Television Service for Schools," *Radio Times*, September 20, 1957, 3.

Stark, James F. 'Anti-Reductionism at the Confluence of Philosophy and Science: Arthur Koestler and the Biological Periphery'. *Notes and Records: The Royal Society Journal of the History of Science* 70:3 (2016): 269–86. https://doi.org/10.1098/rsnr.2016.0021.

Taylor, Charles. *A Secular Age*. Harvard University Press, 2007.

"Television for Schools," *The Times*, December 28, 1956, 7.

Whitehouse, Mary, and Jonny Trunk. *Ban This Filth!: Letters from the Mary Whitehouse Archive*. Faber & Faber, 2012.

CHAPTER 5

Imagining Evolution: Drama and Science Fiction

JOHNSON: He grimaces, my lord, by way of diverting me from certain *fundamental* matters. And indeed I break a pledge! But to say that our ancestors had tails—is it not to make devils of them?

MONBODDO: It is so, sir—if the squirrel and the peacock be arch-fiends, the elephant but a devil in duodecimo, and the Manx cat a blessed celestial visitant.

On first inspection, the BBC radio drama "Strange Intelligence" (1947) seems unremarkable, if a little philosophically highbrow in comparison to most radio output today.[1] Part of the series *Imaginary Conversations* (1946–1956), in which authors and dramatists were invited to create fictional conversations between historical figures, the series was produced by the somewhat forgotten, but nonetheless influential, experimental and (neo)modernist novelist, poet and radio producer Rayner Heppenstall.[2]

[1] "Strange Intelligence", *Imaginary Conversations*, written by Michael Innes, produced by Rayner Heppenstall, BBC Third Programme, June 30, 1947. Excerpt from *Strange Intelligence* by Michael Innes reprinted by permission of Peters Fraser & Dunlop (www.petersfraserdunlop.com) on behalf of the Estate of Michael Innes.

[2] Rayner Heppenstall based the radio series *Imaginary Conversations* on a popular five-volume, nineteenth-century book series of the same name, by the author Walter Savage Landor, published between 1824 and 1829. For more on Rayner Heppenstall's career: Gareth J. Buckell, *Rayner Heppenstall: A Critical Study* (Dalkey Archive Press, 2007).

A. Hall, *Evolution on British Television and Radio*, Palgrave Studies in Science and Popular Culture,
https://doi.org/10.1007/978-3-030-83043-4_5

The novelist and academic J. I. M. Stewart, under his pen name Michael Innes, wrote the episode "Strange Intelligence". It imagined the conversation that may have occurred when the lawyer James Boswell (1740–1795) and the writer Samuel Johnson (1709–1784), stopped at the home of the eccentric Scottish judge Lord Monboddo (1714–1799) on their journey to the Western Isles of Scotland in 1773.[3] Monboddo had a wide range of scholarly interests, including linguistic evolution, and was the progenitor of some proto-evolutionary ideas, such as the development of language skills in primates and early humans.[4] As seen in the epigram of this chapter, "Strange Intelligence" riffed on some of Monboddo's eccentric—and perhaps purely rhetorical—propositions, what in the play Johnson refers to as "his curious doctrine of an ascent of man".

Although included in a subsequent book of the series, it is unlikely that "Strange Intelligence" reached a particularly large audience.[5] Rather, it is included here to serve an illustrative point. Remarkably, this short 30-minute radio drama is one of only a handful of occasions that pre-Darwinian evolutionary ideas and proponents were ever given airtime on BBC radio and television.[6] As "Strange Intelligence" highlights, it is not

[3] Johnson and Boswell featured in an earlier radio show in a similar vein, *Conversations Out of Time* (1933). However, it is likely that for his 1947 script Innes was inspired by the short story "The Monboddo Ape Boy" published by the American author of historical mysteries Lillian de la Torre, as part of her collection of short fictional stories narrated from James Boswell's perspective, *Dr. Sam: Johnson, Detector* (Alfred A. Knopf, 1946).

[4] Stefaan Blancke, 'Lord Monboddo's Ourang-Outang and the Origin and Progress of Language', in *The Evolution of Social Communication in Primates: A Multidisciplinary Approach*, ed. Marco Pina and Nathalie Gontier, Interdisciplinary Evolution Research (Springer International Publishing, 2014), 31–44, https://doi.org/10.1007/978-3-319-02669-5_2.

[5] The series aired on the more culturally highbrow "Third Programme" radio station, which had only launched the previous year in 1946. Only 50% of the population could receive the station and it therefore had substantially smaller audiences than the more established "Home Service" and "Light Programme". The series was never repeated and along with seven others, this episode was published in a limited print run in 1948. *BBC Year-Book 1947* (BBC, 1947), 7; Rayner Heppenstall (Ed.), *Imaginary Conversations* (Secker & Warburg, 1948).

[6] For example, in the period 1923–2009 Charles Darwin's Grandfather, Erasmus Darwin, a proponent of early evolutionary ideas, only featured on approximately 11 broadcasts. Interestingly, the physicist and poet Desmond King-Hele, who wrote a biography of Erasmus Darwin (Charles Scribner's Sons, 1963), wrote four of these broadcasts. Other influential early evolutionary thinkers have featured even less: Comte de Buffon approximately four times, James Hutton four, and even the much debated Jean-Baptiste Lamarck only appears four times in the listings of the *Radio Times*. All figures approximate, taken by searching all

just on extensive historical educational series, such as those produced by the BBC/OU partnership (Chap. 4), where we find a more diverse cast of historical figures. In fact, it is in fictional productions and other non-science broadcasts where, across the BBC's history, we find some of the most arresting content on the more fringe facets of evolutionary science, such as discredited theories and competing ideas. For it is in the spaces where science meets more expansive genres, such as fiction, that we find the most diverse narratives and discussion of evolutionary theory.

Across past issues of the *Radio Times*, content on biological evolution has popped up in some unexpected places; from a tradition of radio dramas on the subject, such as *The Evolution of a Theory* (1948), through to an episode of the long-running primetime chat show *Parkinson*, featuring the zoologist Desmond Morris, which aired in 1977 (Chap. 7).[7] A wide range of departments at the BBC made programmes on evolution in non-scientific contexts, and therefore they are often hard to locate and categorise. They do, however, occur regularly enough to warrant our interest; with dramas alone accounting for approximately 1–2% of the corpus introduced in Chap. 1. Remembering that the corpus only includes shows that contained enough content on biological evolution for the word to be used in their advertised listing, it follows that there have also been a large number of shows that only touched upon the subject in passing, but yet may have still repeated and reinforced a certain framing of the topic.[8]

Despite their often-fragmentary records and the difficulties of categorisation, these non-scientific broadcasts on evolution are important to this study primarily for two reasons. Firstly, their potential to reach new and diverse audiences; chat shows, general issue magazine programmes or dramas will have often reached viewers who had no interest in evolution, and

forms/stylings of names using the BBC Genome Database (https://genome.ch.bbc.co.uk/).

[7] Desmond Morris (b. 1928) is a zoologist, popular author and broadcaster whose 1967 book *The Naked Ape: A Zoologist's Study of the Human Animal*, which proposed evolutionary and sociobiological explanations for many distinct human features, was a worldwide best-seller. From the late 1950s onwards, he regularly featured on British radio and television. For more see Desmond Morris, "Science and Religion: Exploring the Spectrum, Life Story Interviews: Desmond Morris," *National Life Stories*, C1672/16, The British Library, accessed June 14, 2018, https://sounds.bl.uk/Oral-history/Science/021M-C1672X0016XX-0001V0.

[8] For example it is entirely feasible that across the 260 plus episodes of the BBC Religion and Ethics department's Sunday magazine *The Heaven and Earth Show* (1998–2007), evolution occasionally came up in an interview or discussion.

who may not have typically tuned in to general science or documentary broadcasts.[9] Secondly, it is on these programmes that we find most of the content that diverges from contemporary evolutionary science as understood at the time of the broadcasts; whether a focus on pre-Darwinian voices as above, the inclusion of evolutionary stereotypes or tropes or the morally contentious position of science more generally. For example, the most stereotypical framings of the relationship between evolution and traditional Christian beliefs are found in dramas that re-enacted historical Darwinian controversies through the lens of the widely debunked "conflict thesis".[10] While examples like this are found across many areas of non-science broadcasting in the UK, fiction and dramas have done the most to promote and popularise evolutionary themes.

This chapter outlines the development of evolutionary themes in historical dramas and science fiction, the two main genres where evolutionary ideas have featured. In doing so, it considers how a progressive evolutionary narrative became associated with modernist ambitions for broadcasting, and the visions of literary figures George Orwell (1903–1950) and H. G. Wells (1866–1946). I then chart how early science fiction television series, in particular *The Quatermass Experiment* (1953), were built on prior attempts to adapt literary science fiction with evolutionary themes for radio and then television. The tropes and narrative themes used in *Quatermass* carried over into what became the BBC's flagship science fiction television series, *Doctor Who*. In exploring the evolutionary themes of the time-travelling Doctor's adventures, I explore how the series brought evolutionary philosophy and ethics to a mass audience. Following the career trajectory of *Doctor Who* science adviser turned scriptwriter, Kit Pedler, the chapter finishes by analysing the evolutionary themes of the 1970s hit series *Doomwatch*. First, the chapter begins by returning to the

[9] While this is not a particularly new insight, as scholars have long reflected on the influence of science fiction on the public understanding of science, early work in this area often considered science fictions influence secondary or tangential. See, for example, Maurice Goldsmith, *The Science Critic: A Critical Analysis of the Popular Presentation of Science* (Routledge & Kegan Paul, 1986), 59–64.

[10] See, for example, *Horizon: Darwin's Bulldog* (1971), a one-off special, which dramatised the Huxley-Wilberforce debate and the events that led up to this infamous, and historically contested, clash between evolutionary theory and religious dogma. For more on the event and associated debates, see John Hedley Brooke, John Brooke, and Geoffrey Cantor, *Reconstructing Nature: The Engagement Of Science And Religion* (A&C Black, 2000): 35–36; and Bernard Lightman, "Victorian Sciences and Religions: Discordant Harmonies," *Osiris*, Science in Theistic Contexts: Cognitive Dimensions, 16 (2001): 343–66.

earlier tradition of historical radio dramas at the BBC, by introducing an important but to-date overlooked contribution by George Orwell.

Evolutionary Themes in Early Radio Dramas

Although "Strange Intelligence" was one of the first radio dramas to feature evolutionary content, the tradition of radio dramas at the BBC goes back to the institution's inception in the 1920s. While envisioned as a "national theatre of the air" the range of BBC dramas produced in the inter-war period defied such narrow terminology.[11] As well as theatre classics, translations of foreign plays[12] and adaptations of works of literature, the BBC also commissioned a wide range of new "microphone plays"—often innovative and experimental in format.[13] Media historians have explored how the technical and technological challenges of drama via a non-visual medium resulted in inventive solutions, such as the dramatic control panel.[14] While today, the most famous BBC "microphone plays" may be those written by well-known literary figures, such as Dylan Thomas,[15] the majority of content in the early decades was more fleeting.[16] Many broadcasts did not neatly replicate the practice of theatre, and the diversity of BBC radio drama produced was reflected and further reinforced by the separation of Radio Dramas and Radio Features into two departments in the mid-1930s. Despite this institutional split, the boundaries between formats and genres remained fluid, with historically informed series, such as *Imaginary Conversations* produced by the drama department, while in the post-war years non-fiction feature productions increasingly began to use creative devices, such as dramatised re-enactments.

[11] John Drakakis, *British Radio Drama* (Cambridge University Press, 1981), 7–9.

[12] For a popular example from the post-war years see: Jean-Paul Sartre, *Huis Clos*, translated by Marjorie Gabain and Joan Swinstead, produced by Mary Hope Allen. BBC Third Programme, October 4, 1946.

[13] See *BBC Year-Book 1934* (BBC, 1934), 109–13.

[14] See Drakakis, 1–36; and Paddy Scannell, "Features and Social Documentaries" in Paddy Scannell and David Cardiff (eds), *A Social History of British Broadcasting: Volume 1 – 1922–1939, Serving the Nation* (Wiley, 1991), 134–52.

[15] For example, *Under Milk Wood*, 1954.

[16] For an insight into the challenges the BBC faced in soliciting scripts from more established playwrights and dramatists who out of "ignorance and snobbery" considered radio beneath them, see: Louis MacNeice, "Scripts Wanted!", *BBC Year-Book 1947* (BBC, 1947), 25–28.

Orwell, Darwin and the Dramatisation of the Past

The first dramatised radio content about evolution was broadcast in December 1944 when the second episode of *How it was written* (1944–1947), a series that looked at the history of famous books and the biographies of their authors, focused on "The Origin of Species" (Fig. 5.1). The episode used "Darwin's own words" to "help us to understand the drama of the evolution of the idea of evolution".[17] This episode neatly highlights how, despite increasing departmental demarcation, there was still a collaborative culture at the BBC, which shared knowledge and access to experts across subject areas. For while the series was created by radio features writer and producer Stephen Potter, and performed by stage and

Fig. 5.1 Illustration by graphic artist Eric Fraser used to advertise the "The Origin of Species" episode of *How it Was Written* in December 1944. *Radio Times*, December 8, 1944, 16. © *Radio Times*

[17] "The Origin of Species", *How it was written*, produced by Stephen Potter, presented by Cedric Hardwicke, BBC Home Service, December 15, 1944. *Radio Times*, December 8, 1944, 16. All BBC copyright content reproduced courtesy of the British Broadcasting Corporation. All rights reserved.

Hollywood actor Cedric Hardwicke, the episode on Darwin's *Origins* was based on factual content provided by the naturalist and broadcaster James Fisher.[18]

The most notable early radio drama to feature evolution was an episode of *Voyages of Discovery*, aired in March 1946, which focused on the voyage of the *Beagle* and its foundational role in Darwin's career.[19] As with *Imaginary Conversations*, Rayner Heppenstall produced the series, with the episode on the *Beagle* employing the literary talents of Heppenstall's long-time friend; noted journalist and author George Orwell.[20] In the dramatisation, "the intellectual struggle and the physical adventure" were "woven" together by Orwell, using Darwin's recently published letters and notebooks to inform the dialogue.[21] Beginning with a narrated précis of Darwin's career and impact on society, the audience heard a brief re-enactment of the infamous clash at the British Association meeting in Oxford where "Darwin was fiercely attacked, and equally strongly defended by his disciples".[22] By opening with this scene before heading back in time to the action aboard the *Beagle*, Orwell let the audience know from the off that the story he wished to tell was one centred on conflict.

After setting up this conflict-centred narrative frame, Orwell introduced the episode's antagonist: the captain of the *Beagle*, Robert Fitzroy,

[18] *On the Origin of Species* was the only scientific book to feature in this short-lived series, other books featured included Flaubert's *Madame Bovary*, and Dickens' *The Pickwick Papers*. The format was occasionally resurrected under different names in the decade that followed, for example see *The Making of a Novel*, written and presented by Robert Speaight, BBC Third Programme, January 23, 1951. The episode of *How it was written* on *Origins* was later reworked and expanded by Potter, becoming 1948s *The Evolution of a Theory*. The content provided by Fisher remained, but the dramatic approach was expanded producing a two-person duologue between Darwin and his close friend the botanist and explorer Joseph Hooker.

[19] "The Voyage of the 'Beagle'", *Voyages of Discovery*, produced by Rayner Heppenstall, written by George Orwell, BBC Home Service, March 29, 1946. Full script: "2953. The Voyage of the 'Beagle'", in George Orwell, *Smothered Under Journalism 1946* (Secker & Warburg, 2001), 179–201.

[20] In 1947 Heppenstall also produced the first-ever radio performance of *Animal* Farm, which Orwell himself adapted for the wireless: *Animal Farm,* written by George Orwell, produced by Rayner Heppenstall, BBC Third Programme, January 14, 1947. For more on Heppenstall's relationship with Orwell see: Rayner Heppenstall, *Four Absentees* (Barrie and Rockliff, 1960).

[21] George Orwell, "The Voyage of the 'Beagle'", *Radio Times*, March 22, 1946, 4.

[22] Orwell, *Smothered Under Journalism 1946*, 182.

portrayed as a fundamentalist who believed in a literal interpretation of scripture. As the *Beagle* wound its way around South America, Orwell repeatedly contrasted Fitzroy's fixed position against Darwin's inquisitive and open scientific approach. While this tension and the wider script in general was a largely accurate account resembling the views recorded in Darwin's personal correspondence, Orwell's decision to focus the play solely on this tension was a conscious choice. Orwell used Fitzroy to personify the wider societal anger, debates and reflections on the place of man on Earth that the publication of *Origins* caused in the second half of the nineteenth century.

Throughout the play, Orwell alternated between third- and first-person narrative modes; introducing the voice of Darwin's widow in addition to the traditional narrator, to add an extra-temporal point of view to the narrative. Although Emma Darwin herself quipped that this was merely "to give the narrator a rest", the unusual device allowed the audience to discern between Darwin's views as a young man, the general societal position following his death in 1882 and the contemporary view as the audience would have understood it in 1946. Often influenced—in this case directly—by emergent literary and theatrical trends, as the development of genres abounded during the second half of the twentieth century, radio and television experimented with many narrative techniques from multiple narrators, to breaking the fourth wall and simultaneous multiple plot lines.[23]

As we will see in Chap. 8, the conflict-centred frame adopted by Orwell, with its focus on the personal travails and consternation that Darwin's ideas caused to his own mental well-being and religious piety, became the most common feature of historical flashbacks and re-enactment segments on future programmes. Orwell's play skilfully utilised storytelling devices to make it clear that the conflict between Darwin and Fitzory, and by extension Darwinism and British society, was about fundamentalist tradition versus open natural enquiry. Rather than reducing the conflict to religion versus science, as a dramatist, Orwell was more concerned with exploring how Darwin's ideas may have shook foundational Enlightenment

[23] For an introduction to the most common narrative approaches in television see Kristin Thompson, *Storytelling in Film and Television* (Harvard University Press, 2003). For a case-study that outlines the development of more complex narrative forms in US television during the 1990s see Jason Mittell, "Narrative Complexity in Contemporary American Television", *The Velvet Light Trap* 58:1 (2006): 29–40, https://doi.org/10.1353/vlt.2006.0032.

beliefs that held humanity in a central manifest destiny. However, in future productions as flashbacks to Darwin were often reduced to just 30-second segments of longer science broadcasts, or as a device to contextualise current debates such as creationism (Chap. 7), the nuance of the conflict explored in Orwell's play was most often lost. Subsequent, less in-depth dramatisations most commonly reduced the narrative to a conflict frame centring on Darwin's theory clashing with religion per se, rather than with religious authority, or specific areas of religious doctrine. To borrow from social psychological studies on evolution and religion, those producing and writing these shows often "projected" contemporary ideas back onto historical figures. In particular, their own "in-group" assumption that evolution was not compatible with traditional religious beliefs was cast back onto earlier debates and religious groups, reducing complex historical deliberations to fit convenient media frames familiar to audiences.[24]

As with the development of the illustrated talk format for school broadcasts (Chap. 4), many of the techniques developed for BBC radio dramas prior to the Second World War were creative solutions to make non-visual, and complex content accessible and arresting for a general audience. As the BBC continued to grow, much of the tacit technical know-how, such as the use of the dramatic control panel to deliver live sound effects, and some of the successful narrative devices, such as stream of consciousness narration, became embedded into the technologies, departmental processes and radio formats themselves.[25] The most successful became standard radio practice, with many, such as the historical dramatisation eventually transitioning to television.

Wells, Ethics and Dystopian Futures

Alongside its rapid popularisation and the blossoming of new formats, radio was also central to modernist ideals of heavyweight thinkers and

[24] For an introduction to these issues and concepts see: Fern Elsdon-Baker, "Creating hard-line 'secular' evolutionists: The Influence of Question Design on Our Understanding of Public Perceptions of Clash Narratives", and Carissa Sharp and Carola Leicht, "Beyond Belief Systems: Promoting a Social Identity Approach to the Study of Science and Religion", both in Fern Elsdon-Baker and Bernard Lightman (Eds.), *Identity in a Secular Age: Science, Religion, and Public Perception* (University of Pittsburgh Press, 2020).

[25] For more on some of these early approaches and technologies see: Angela Frattarola, 'The Modernist "Microphone Play": Listening in the Dark to the BBC', *Modern Drama* (2010) https://doi.org/10.3138/md.52.4.449.

public intellectuals of the 1930s. Modernist literary figures, such as E. M. Forster and T. S. Elliot contributed radio essays, dramatisations, radio features and commentary on the role and potential of radio for improving society.[26] Among the cast of early promoters, the writer H. G. Wells stands out with regard to the integration of science, both in his writing and radio commentaries. In the 1880s Wells had studied biology under Thomas Henry Huxley and was greatly influenced by his thinking, infusing much of his writing, both fiction and non-fiction, with ideas on the fundamental unity of all life on earth.[27] Wells' writing often connected evolutionary, social and ethical themes, and in doing so drew directly from T. H. Huxley's work on evolutionary theory and its ethical implications. As the literary scholar, Todd Avery summarises:

> No British writer of the early twentieth century had more thoroughly absorbed Huxley's ethical views or promoted them more vigorously than H. G. Wells. Little of Wells work, from the scientific romances of the 1890s to even the pessimistic reflections on society and politics in *Mind at the End of Its Tether* in the 1940s, can be read without noticing the profound and lasting impact of Huxley's ethical ideas.[28]

Not only was Wells strongly influenced by T. H. Huxley, during the 1920s and 1930s he was also part of the same centre-left political and social milieu as Julian Huxley, contributing articles to the short-lived *Realist* magazine (Chap. 2) and co-authoring the best-selling three-volume textbook *The Science of Life* (1929) with his son P. G. Wells and Huxley (Chap. 4). Wells' philosophy diverged in significant ways from Huxley's scientific humanism, most notably with regard to the importance of religious experience, and the role of the individual. However, both of their visions for the future were built on evolutionary ideas and promoted science as central to humanity's progress.[29] As detailed by the historian Peter Bowler, Wells' was among a group of writers in the first half of the twentieth

[26] Todd Avery, *Radio Modernism: Literature, Ethics, and the BBC, 1922–1938* (Ashgate Publishing, Ltd., 2006), 61–67 and 111–136.

[27] John Batchelor, *H. G. Wells* (Cambridge: Cambridge University Press, 1985), 164; and David C. Smith, *H. G. Wells, Desperately Mortal* (Yale University Press, 1986), 10–13.

[28] Avery, *Radio Modernism*, 90.

[29] Robert Bud, '"The Spark Gap Is Mightier than the Pen": The Promotion of an Ideology of Science in the Early 1930s', *Journal of Political Ideologies* 22:2 (2017): 169–81, https://doi.org/10.1080/13569317.2017.1298548, 171–3.

century whose output spanned science fiction, popular science writing, futurology and ethics/philosophy of science.[30] In producing popular factual books like *The Science of Life* alongside his science fiction, Wells did more than just seek authority for the accuracy and plausibility of the science depicted in his fictional adventures. He utilised different narrative formats and genres to reinforce and popularise his particular vision for the future of humanity.[31]

As we will see in this chapter, a blurred boundary between fiction and non-fiction remained central to British broadcasts on evolution. When we are considering the effect that popular media may have on shaping someone's worldview on evolution, or any scientific subject, this blurred boundary highlights the importance of incorporating fictional accounts into our analysis. It also highlights why in this chapter I broaden the scope beyond traditional evolutionary fictions, such as Wells' *The Time Machine* (1895), to stories that while not overtly Darwinian contain narratives that wrestle with evolutionary ethics, and the implications that man's knowledge of evolutionary science place on humanity's actions.[32]

As radio continued its rise in popularity, alongside more diverse drama features and plays, during the 1930s the BBC continued to expand the scope of its other literary content such as dramatic readings and book review shows.[33] In doing so, they expanded from plain recitals of Wells' more traditional novels, such as the rags-to-riches social commentary of *Kipps* (1905), first broadcast in 1926, to his earlier works of science fiction, such as *The First Men in the Moon* (1901) and *The Time Machine* (1895), first broadcast respectively in 1935 and 1940.[34] One of the first

[30] Peter J. Bowler, *A History of the Future: Prophets of Progress from H. G. Wells to Isaac Asimov* (Cambridge University Press, 2017).

[31] Peter Bowler, 'Parallel Prophecies: Science Fiction and Futurology in the Twentieth Century', *Osiris* 34:1 (2019): 121–38, https://doi.org/10.1086/703952.

[32] Mark Bould and Sherryl Vint, *The Routledge Concise History of Science Fiction* (Routledge, 2011), 31–32.

[33] Drakakis, *British Radio Drama*, 7–12.

[34] Reflecting the wider trajectory of the seriousness afforded to science fiction, nearly all of these early readings were part of the *For the Schools* slot on the BBC Home Service. See Smith, *H. G. Wells, Desperately Mortal*, 511 for a full list of H. G. Wells' BBC broadcasts. "Kipps", written by H. G. Wells, March 25–31, and July 1, 1926, various BBC regional stations; "English Literature, 'The First Men in the Moon'", written by H. G. Wells and read by Colin Milne, November 1935–August 1936, BBC Regional Programme; and "Dramatic reading: 'The Time Machine', by H.G. Wells", *For the Schools*, adapted for broadcasting by Douglas R. Allan, January–August 1940, BBC Home Service.

literary works to popularise the concept of time travel, The Time Machine, explored concepts such as evolutionary determinism and social and biological degeneracy.[35] It proved to be a popular BBC go-to, with the 1935–1936 reading followed by dramatic adaptations in 1940 and 1947, and as discussed below, a television adaptation as early as 1949.[36]

One of Wells' most overt evolutionary fictions, *The Island of Dr Moreau* (1896), in which a shipwrecked protagonist found himself on an island where a disgraced physiologist vivisected animals to create human-like "Beast Folk", did not receive the BBC dramatisation treatment until 1990.[37] However, the enduring allure of Wells' narrative featuring the now stereotypical depiction of a "mad scientist" amorally interfering with nature is demonstrated by the fact the novel has been subject to three Hollywood adaptations—*Island of Lost Souls* (1932), The Island of Dr. Moreau (1977) and The Island of Dr. Moreau (1996). Wells published the original novel as discussion of Francis Galton's concept of eugenics was reverberating across British society, and as such, the storyline explored the ethical implications of attempts to alter our own heritable material. [38] The story presented the first instance of the science fiction convention of "uplift", whereby a developmental process is used to transform a species into a more intelligent form.[39] While our scientific and technological

[35] See respectively: Kirby Farrell, "Wells and Neotony", in George Edgar Slusser, Patrick Parrinder and Danièle Chatelain (Eds), *H.G. Wells Perennial Time Machine* (University of Georgia Press, 2001), 65–75; and Richard Barnett, 'Education or Degeneration: E. Ray Lankester, H. G. Wells and The Outline of History', *Studies in History and Philosophy of Science Part C: Studies in History and Philosophy of Biological and Biomedical Sciences* 37:2 (2006): 203–29, https://doi.org/10.1016/j.shpsc.2006.03.002.

[36] "Senior English. Dramatic reading: 'The Time Machine', by H. G. Wells", *For the Schools*, adapted by Douglas R. Allan, May 1940, BBC Home Service; "Senior English Serial: 'The Time Machine,' by H. G. Wells," *For the* Schools, adapted by Penelope Knox, March 1947, BBC Home Service; and "H.G. Wells' The Time Machine", adapted and produced by Robert Barr, January 25, 1949, BBC Television. As with most *For the Schools* content during the period, the radio versions of *The Time Machine* were repeated several times, for example the Penelope Knox adaptation was repeated again in November/December 1947.

[37] "Saturday-Night Theatre: The Island of Dr Moreau," by H. G. Wells, dramatised by David Calcutt, September 1, 1990, BBC Radio 4.

[38] David A. Kirby, 'Are We Not Men?: The Horror of Eugenics in The Island of Dr. Moreau', *ParaDoxa* 17 (2002), 93–108; David A. Kirby, 'The Devil in Our DNA: A Brief History of Eugenics in Science Fiction Films', *Literature and Medicine* 26:1 (2007): 83–108, https://doi.org/10.1353/lm.2008.0006.

[39] For a timeline of works of science fiction that utilise uplift in their plots, see: "Uplift (science fiction)", last modified June 14, 2021, https://en.wikipedia.org/wiki/Uplift_(science_fiction).

capabilities concerning eugenics and genetic manipulation changed vastly in the decades following Wells' novel, our fears about the ethics of interfering in evolutionary processes, and whether we can reduce all of the human condition to our genetics, persisted. As the science and technology studies scholar David Kirby summarises:

> Eugenics is about transformation, and this is why filmmakers keep coming back to eugenic themes. Cinematic attacks on eugenics are never directed at the mechanism of hereditary manipulation. The technologies on the screen, whether potion-filled test tubes, electronic devices, or the actual tools of molecular biology, are visual cues that signify genetic transformation. In science fiction cinema, however, transformations never match the eugenic dream of a superior humanity. Rather, these films feature graphic transformations in which eugenic plans turn human beings into monsters.[40]

Even in the most fantastical depictions, scientific accuracy—or at least plausibility—remains important.[41] When making the *Island of Lost Souls* (1932), the filmmakers even invited Julian Huxley onto set to get his endorsement for the accuracy of the eugenic science depicted in the movie.

Perhaps the most infamous science fiction broadcast during radio's early decades is Orson Welles' adaptation of H. G. Wells' *The War of the Worlds*, broadcast in the United States on CBS radio in 1938. Despite reaching a sizeable audience, estimated to be between 2 and 4 million, initial accounts grossly exaggerated the scale of nationwide mass panic caused by the realism of the broadcast's Martian invasion.[42] Rather than simply highlighting that society fears alien invasion, media studies scholar Jeffrey Sconce has argued that the huge initial press coverage the broadcast generated and the exaggeration and persistence of the mass-hysteria myth highlight our continued collective anxieties regarding the role of mass communication.[43] *The War of the Worlds* first received airtime in the UK during the last year of the war in the spring of 1945, and was adapted

[40] Kirby, 'The Devil in Our DNA', 105.

[41] David A. Kirby, *Lab Coats in Hollywood: Science, Scientists, and Cinema* (MIT Press, 2011), 34–5.

[42] A. Brad Schwartz, *Broadcast Hysteria: Orson Welles's War of the Worlds and the Art of Fake News* (Farrar, Straus and Giroux, 2015).

[43] Jeffrey Sconce, *Haunted Media: Electronic Presence from Telegraphy to Television* (Duke University Press, 2000), 114–118.

into a dramatised radio serial by the prolific author and screenwriter Jon Manchip White, broadcast in the summer of 1950.[44]

As these initial adaptations of H. G. Wells' works attest, radio and the emergent genre of science fiction turned out to be great bedfellows. Science fiction storylines, with their focus on errant technologies and dystopian futures, mirrored well the anxieties that mass broadcast technology itself wrought upon the public. Further, the evolutionary themes at the heart of Wells' science fiction—in particular, the idea of human interference with natural processes and the ethics of evolutionary science—proved to be enduring leitmotifs of BBC produced science fiction. While science fiction on the radio continued to be popular, often providing a successful testing ground for new voices in the genre,[45] it was television, with its visual potential, that would go on to produce some of the most memorable depictions of evolutionary themes in hit primetime shows.

Evolutionary Themes in British Science Fiction Television

Science fiction was there at the very beginning of experimental television broadcasts in the UK.[46] In February 1938, the BBC broadcast the first-ever televised science fiction, a short adaptation of Karel Čapek's play, *R.U.R.* (*Rossum's Universal Robots*, 1920), in which artificial people, the *roboti* forced to work in factories producing human goods, overthrow and massacre their human masters. [47] The play's fantastical plot gave the production team the opportunity to experiment with lots of technical aspects, from testing which costumes worked best at such low resolution,

[44] Smith, *H. G. Wells, Desperately Mortal*, 511; *The War of the Worlds*, by H. G. Wells, adapted Jon Manchip White, May–July 1950, BBC Light Programme, and September–October 1950, BBC Home Service.

[45] Perhaps the most famous example is *The Hitchhiker's Guide to the Galaxy* by Douglas Adams, the original BBC 4 radio comedy series broadcast in 1978, pre-dates the novel (1979), the BBC television series (1981), the computer game (1984) and the more recent feature film (2005).

[46] Derek Johnston, 'Experimental Moments: R.U.R. and the Birth of British Television Science Fiction', *Science Fiction Film & Television* 2:2 (2009), 254.

[47] *R.U.R. (Rossum's Universal Robots)*, by Karel Čapek, adapted by and produced by Jan Bussell, February 11, 1938, BBC Television; and Mark Bould, "Science Fiction Television in the United Kingdom," in J. P. Telotte, *The Essential Science Fiction Television Reader* (University Press of Kentucky, 2008), 210. Čapek's play introduced and popularised the word robot, from the Czech word *robota*, which means forced labour.

to trialling new superimposition techniques to multiply the size of the robot army. Initially envisaged as a medium suited to relaying live events, in these early drama broadcasts, BBC producers were still experimenting with the form; attempting to show TV drama could offer something distinct from radio, theatre and film.[48]

Although analysis of *R.U.R* often centres on its critique of capitalism, at its heart the play is an exploration of the nature versus culture divide, which was so central to emergent literary science fiction in the late nineteenth century. Like Wells' "Beast folk" and the creations of Shelley's Dr Frankenstein, Čapek's *roboti* are not mechanical, but rather are "uplifted" living beings assembled from biological components made in factories by humans. Thus, while geopolitical commentary was front and centre, *R.U.R.* also more subtly explored colonial constructions of race and their associated evolutionary ethics, a theme also central to early science fiction novels by authors such as Wells.[49]

With television suspended during the Second World War, it was not until the late 1940s that experiments with science fiction television (SFTV) continued and British SFTV began to develop in earnest. Again, the BBC turned to *R.U.R* and producer Jan Bussell; this time the full 90-minute version of the play aired in March 1948 almost exactly a decade after the first.[50] Among the cast of the 1948 version was Patrick Troughton, a young actor who would later become known for his work in fantasy and science fiction roles, most notably as the second incarnation of the Doctor in *Doctor Who*.[51] BBC Television followed up *R.U.R.* with an adaptation of H. G. Wells' *The Time Machine* broadcast in January 1949.[52] The one-hour live teleplay was adapted and produced by former war correspondent and documentary producer Robert Barr (1909–1999), working closely with designer Barry Learoyd (1915–2003). In promoting the teleplay, the

[48] Johnston, 'Experimental Moments'.

[49] John Rieder, *Colonialism and the Emergence of Science Fiction* (Wesleyan University Press, 2012), 97–122.

[50] No full scripts for either the 1938 or 1948 versions of *R.U.R.* are held at the BBC Written Archives, with only one small folder covering both productions, see: "*R.U.R.* 1938 and 1948", T5/443, BBC-WA.

[51] *R.U.R. (Rossum's Universal Robots)*, by Karel Čapek, adapted by and produced by Jan Bussell, March 4 & 5, 1938, BBC Television; "R.U.R. (Rossum's Universal Robots)", *Radio Times*, February 27, 1948, 25 & 27.

[52] *H.G. Wells' The Time Machine*, by H. G. Wells, adapted and produced by Robert Barr, January 25, 1949, BBC Television.

producers focused on their ingenious solutions to the visual challenges of a story that covered over 800,000 years. Yet in the limited audience responses we have for the production—through letters in the *Radio Times*—the audience was not appreciative of the visual experimentation, but rather more concerned with the length, comprehensibility and their enjoyment of the broadcast.[53]

Kneale and Literary British Science Fiction

Both of these early post-war attempts at televising science fiction were adaptations of already existing stories. Like with the expansion of educational broadcasting (Chap. 4), it was not until the 1950s and the threat posed by the launch of commercial television that BBC TV began to commission its own dramas. Despite the increased technical capabilities of the new purpose-built television studios at Lime Grove, the focus of early BBC TV dramas was on dialogue and intimacy of setting. The first notable commissioned drama to go against this trend was *The Quatermass Experiment* (1953),[54] at the time described as a "thriller", but today recognised as a seminal SFTV series.[55] The series, performed live on a sparse set, reflected austere post-war British society, and throughout the plot was infused with the techno-anxieties of the early Cold War. The action followed the first-manned mission into space led by the archetypal protagonist scientist, the single-minded and moral Professor Quatermass. Upon

[53] John Swift, "To the World's End in Sixty Minutes", *Radio Times*, January 21, 1949, 25; and "Viewers are Saying", *Radio Times*, February 4, 1949, 25. As discussed in, Derek Johnston, "Genre, Taste and the BBC: The Origins of British Television Science Fiction" (University of East Anglia, PhD Thesis, 2009), 55–56. For more on other early television science fiction adaptations see Johnston, "Genre, Taste and the BBC", 71–109.

[54] *The Quatermass Experiment*, written by Nigel Kneale, produced by Rudolph Cartier, July–August 1953, BBC Television. Interested readers can watch the opening two episodes of *The Quatermass Experiment* online at the following link: https://archive.org/details/TheQuatermassExperiment-Incomplete *(last accessed 18/08/2021)*. Unfortunately, only the opening two episodes were recorded and as the show was performed and broadcast live no video or audio of the following four episodes remains. All six episodes scripts are held on microfilm at the BBC-WA.

[55] Catherine Johnson, "Exploiting the Intimate Screen: The Quatermass Experiment, Fantasy and the aesthetic potential of early television drama", in Janet Thumim (Ed.), *Small Screens, Big Ideas: Television in the 1950s* (Bloomsbury Academic, 2002), 181–194; Mark Bould, "Science Fiction Television in the United Kingdom", 210–11; Johnston, 'Experimental Moments', 254.

crash-landing in London, the rocket was found to contain only one crew member, Victor Carroon. It transpired that while in space aliens visited the rocket, and Carroon, who seemed to have absorbed the consciousness of his missing colleagues, slowly began to mutate into a plant-like alien organism. Again evolutionary ideas and ethics were central to the plot; when Quatermass discovered that if the organism began to spore it would destroy life on Earth, he used the remnants of the crew's consciousness—their humanity—now incorporated into the alien, to defeat the organism before its apocalyptic proliferation. The subsequent film adaptation of the series, *The Quatermass Xperiment* (1955) took the hybrid body idea even further, when the mutating Victor Carroon smashed his hand into an ornamental cactus in the hospital resulting in a cactus-arm.[56]

In a period when mass primetime Saturday night television viewing was still being established, *The Quatermass Experiment* was a popular and critical success.[57] Written by BBC TVs first full-time scriptwriter Nigel Kneale (1922–2006), the series was an attempt to push television beyond the influence of radio and theatre, to incorporate more visual and cinematic styles.[58] Kneale and series' producer Rudolph Cartier (1904–1994) deemed a fantastical setting the best vehicle for this and in doing so, "they attempt[ed] to expand the scope of television drama into the spectacle afforded by science fiction narratives, while simultaneously recognizing and exploiting the familiarity and intimacy created through the transmission of an extended serial narrative into the viewer's domestic space".[59] This dual quality of science fiction narratives, their ability to create fantastical expansive worlds, while also building relatable protagonists via a serial format, had first been perfected in early serialised literary works, such as Wells' *Time Machine* originally published over several editions of *The New Review* in 1895. By transferring this format to the small screen, and creating a distinctly British and adult feel for the series, Kneale helped to connect British SFTV to a literary tradition perceived as more highbrow by

[56] Highlighting the influence of *Quatermass* on subsequent SFTV, this hybrid human-cactus was later referenced in the 1980 *Doctor Who* story "Meglos".

[57] See: Radio Critic, "Saturday Serial", *The Manchester* Guardian, August 5, 1953, 3; C. A. Lejeune, "Television Notes", *The Observer*, July 26, 1953, 11; Gilbert Harding, "Gilbert Harding's Notebook", *Picture Post*, November 14, 1953, 40; D. Duke, "Like Carroon", *Radio Times*, September 4, 1953, 9; and P. Thomas, "Thwarted" *Radio Times*, September 4, 1953, 9.

[58] Catherine Johnson, "Exploiting the Intimate Screen", 183–84.

[59] Catherine Johnson, "Exploiting the Intimate Screen", 193.

British elites than the American pulp fiction that was popular among British children and teenagers.[60] The elevation of science fiction in the cultural hierarchy of British society was neatly encapsulated in the listing for a 1960 radio show, *The Vanishing Pulps* in which the art critic Lawrence Alloway reported on "English science-fiction":

> Mr. Alloway discusses the recent development of a trend towards a more 'scientific' kind of English science-fiction reflected in the dropping of the word Astounding' from the title of one of the leading SF magazines. He sees this distaste for pulp magazine serials as a symptom of sociological change.[61]

In bypassing tensions around the Americanisation of British society, Kneale and Cartier's *Quatermass* broadened BBC senior figures' understanding of what constituted science fiction, and thus was a significant contributor to this "sociological change".

Although subjects lifted from earlier literary science fiction and explorations of evolutionary ethics remained popular over the next two decades, the BBC developed their own distinct brand of televised science fiction. As the film and literature scholar Mark Bould summarises, the British SFTV approach was shaped by "its development within a national context of imperial decline", which contrasted sharply with a US narrative of ascendancy during the inter-war and post-war decades. As a result, "British SFTV has often evoked these anxieties about the end of empire and the concomitant decentring of the United Kingdom within a globalised economy—while celebrating the erosion of traditional privilege and authority in the world of consumer capital—by foregrounding changing class relationships and gender roles".[62]

On the back of the success of *Quatermass*, a small but steady stream of BBC commissioned science fiction dramas followed, including the successful sequels *Quatermass II* (1955) and *Quatermass and the Pit* (1958–1959), both written by Kneale.[63] Alongside these dramas, the BBC

[60] Derek Johnston, "The BBC "Versus" Science Fiction: The Collision of Transnational Genre and National Identity in Television of the Early 1950s", in Tobias Hochscherf, James Leggott, and Donald E. Palumbo (Eds.), *British Science Fiction Film and Television: Critical Essays* (McFarland, 2014), 40–49.

[61] "The Vanishing Pulps", *Radio Times*, December 1, 1960, 29.

[62] Mark Bould, "Science Fiction Television in the United Kingdom", 209–230.

[63] A film version of the original series, *The Quatermass Xperiment* produced by UK Hammer Horror Productions and adapted for the screen by writer and director Val Guest was released in cinemas in 1955.

capitalised on the growing popularity of the genre, particularly among the younger generation, featuring science fiction on book and film review shows, as well as commissioning new features to discuss aspects of the genre and its most famous authors.[64] Features producers also capitalised on the popularity of science fiction to reach wider audiences with science content, with one-off broadcasts, such as *The Fiction of Science* (1957), *Monsters in Miniature* (1957), and *Frontiers of Science: Space Travel* (1957).[65] In doing so, the BBC cemented science fiction's place on British radio & TV, ensuring that the British public did not just see it as light entertainment, but as an art form that could generate serious reflection and discussion about the role of science and associated new technologies in society.

Newman and the Rise of the Serial

Despite early successes with the *Quatermass* serials, by the early 1960s BBC Television's audience figures were significantly lagging behind new commercial upstart ITV. As part of moves to regain ground and focus more resources on television, in 1963 the BBC hired Sydney Newman, a successful Canadian drama producer working for the ITV regional franchise, ABC.[66] Newman gave serials an equal footing with individual plays, and in restructuring the BBC Drama Group, reinvigorated their protracted search for a new science fiction series. What followed was the creation of a science fiction serial by committee. The Drama Group knew the early Saturday evening time slot they needed to fill, they had a shortlist of suit-

[64] See respectively: an episode of *The Younger Generation: Under 20 Review* (BBC Light Programme), aired on April 8, 1953 on which two newly published SF collections were discussed; an episode of *Film Time* (BBC Home Service) from November 29, 1951 which reviewed new SF movie releases, including *The Thing from Another World* (1951); and the one-off special *Jules Verne and Science Fiction* (January 11, 1955, BBC Home Service) on which the novelist and playwright Clemence Dane discussed her love of Verne and his influence on current SF trends.

[65] *The Fiction of Science*, by John H. Steele, June 3, 1957; *Monsters in Miniature*, by Kenneth Bisset, February 15, 1957, BBC Home Service; and *Frontiers of Science: Space Travel*, produced by James McCloy, October 11, 1957, BBC Television.

[66] Early successes for Newman at ABC included the space travel focused *Target Luna* (1960), and its sequels *Pathfinders in Space* (1960), *Pathfinders to Mars* (1960–61) and *Pathfinders to Venus* (1961). See Mark Bould, "Science Fiction Television in the United Kingdom", 213.

able recent science fiction novels and they developed character profiles accordingly to match up with each member of the target family demographic they hoped to reach. Despite being beset by production issues and being overshadowed by the assassination of President Kennedy the day before, on Saturday November 23, 1963, the BBC aired "An Unearthly Child", the first episode of the now longest-running science fiction television series *Doctor Who* (1963–1989, 2005–pres.).[67] At its core, the concept was simple. Newman wanted short SF serials linked by a few core characters,[68] so the series followed the Doctor, a mysterious "Time Lord", who with his human companions travelled through time and space in his stolen time machine, the TARDIS, thwarting evil adversaries.[69] Each short-run serial of several episodes was a self-contained story, usually created by one writer.[70]

Although upon first viewing the series is not what we might think of as typical evolutionary fiction, due to the centrality of time travel, in most seasons, evolutionary themes were never far from the surface. [71] The "scientific but whimsical" Doctor and his time machine were openly influenced by *The Time Machine*, with knowing nods to H. G. Wells regularly featuring across the seasons.[72] Covering a similarly long-time period as *The Time Machine*, the evolution of species and human culture was a recurring theme. At the end of the very first episode, the Doctor, his original com-

[67] The production was beset with problems from the outset, scripts were dropped, technical issues abounded and the first episode had to be re-recorded to soften the characterisation of the Doctor. "An Unearthly Child", *Doctor Who*, written by Anthony Coburn, November 23, 1963, BBC Television, available online: https://archive.org/details/doctorwhoanunearthlychildepisode1incolor_202003 *(last accessed 21/06/2021)*; James Chapman, 'Fifty Years in the TARDIS: The Historical Moments of Doctor Who', *Critical Studies in Television* 9:1 (2014): 44–47; and Lindy A. Orthia, 'Antirationalist Critique or Fifth Column of Scientism? Challenges from Doctor Who to the Mad Scientist Trope', *Public Understanding of Science* 20:4 (2011): 525

[68] Chapman, 44.

[69] TARDIS stands for "Time and Relative Dimension in Space", while other TARDISes can take on different forms, due to a broken circuit the Doctor's is always, the now iconic, blue police phone box.

[70] The first season was composed of eight serials, written by six different authors. Serials ranged in length from two to seven episodes.

[71] Mark Bould and Sherryl Vint, *The Routledge Concise History of Science Fiction* (Routledge, 2011), 31–32.

[72] Like the Doctor, the protagonist in Wells' *The Time Machine* is known simply as the *Time Traveller*. John Tulloch and Manuel Alvarado, *Doctor Who: The Unfolding Text* (St. Martin's Press, 1983), 121–127, http://archive.org/details/doctorwhounfoldi00tull.

panion Susan and her teachers went back in time 10,000 years. Then in the second episode, "The Cave of Skulls" they were captured by stone-age early *Homo sapiens.* The tribe, who were struggling to recreate fire when the time-travellers arrived, were depicted as stereotypical cave-dwellers,[73] with Neanderthal-like features—stocky, hunched, club-wielding, human-sacrificing primitive hominids who lack the humanity of modern humans.[74]

The opening serial of autumn 1963 received an average of six million viewers over its four episodes; however, this represented only 12.3% of the potential audience.[75] Harking back to the successes of early radio dramas, such as Orwell's episode of *Voyages of Discovery*, the original plan for *Doctor Who* was to alternate between historical adventures, like this first serial and those set in the future or on alien planets and locations. It was the alien-focused second serial, "The Daleks", which went from an audience of 6.9 million in episode one to 10.4 million by the seventh episode, which proved to be the big breakthrough for the series.[76] It also alluded to a pattern: using independent writers for each serial, resulted in fluctuating viewing figures that troubled the show throughout the rest of the decade.

It was during one of these slight wanes in popularity in 1965–1966 that a new production team was brought in. The incoming producer Innes Lloyd phased out the increasingly inaccurate historical narratives, and with the help of new script editor Gerry Davis brought in a scientific advisor, senior lecturer in ophthalmology at the University of London, Kit Pedler

[73] "An Unearthly Child", 21:50; "The Cave of Skulls", *Doctor Who*, written by Anthony Coburn, November 30, 1963, BBC Television. For more on the stereotypical depiction of early hominids and Neanderthals, including their representation in H. G. Wells' *Outline of History* (1920), see Amanda Rees and Charlotte Sleigh, *Human* (Reaktion Books, 2020), 54–67.

[74] The most extensive lifting from H. G. Wells in *Doctor Who* appeared in the 1985 serial *Timelash*, in which Wells himself featured, encountering various phenomena supposed to have inspired his novels, and in which, in a nod to *Dr Moreau*, the arch antagonist was a humanoid-animal hybrid. Glen McCoy, *Dr Who: Timelash* (W. H. Allen); and "Timelash" *Doctor Who in Vision*, Issue 83, March 1999.

[75] Chapman, 'Fifty Years in the TARDIS', 47.

[76] Written by screenwriter Terry Nation, who would go on to create seminal BBC SFTV series *Survivors* (1975–77) and *Blake's 7* (1978–81), the second serial introduced the now ubiquitous aliens the Daleks, with their sinister order to "Exterminate!" an echo of the cry "Finished" shouted offstage by the robots in Čapek's *R.U.R.* Johnston, 'Experimental Moments', 261; and the Doctor Who Guide online, https://guide.doctorwhonews.net/info.php?detail=ratings&type=date&order=asc *(last accessed 18/06/21)*, where interested readers can find viewing figures for all episodes of the series.

(1927–1981). Lloyd brought Pedler in not to vet the scientific accuracy of the show, but rather to replace many of *Doctor Who's* fantasy elements with more plausible science fiction, based on real-life cutting-edge science and technology. [77]

Pedler and Techno-Scientific Anxieties

Pedler had first come to the attention of the BBC in 1965 when the science and technology show *Tomorrow's World* (1965–2003) visited his university lab.[78] Lloyd selected Pedler for the advisory role on *Doctor Who* ahead of several more prominent public scientists, including the physician and novelist Dr Alex Comfort. Pedler was a science fiction fan, and deemed to be more "flexible" when it came to understanding how science was utilised in dramatic interpretations. Remarkably, after contributing the central idea of a rogue sentient computer to season three serial, "The War machines" (1966), Pedler, who had no prior experience of writing fiction, was commissioned to write a serial for season four. [79] "The Tenth Planet", which ended up being co-written with script editor Gerry Davis, introduced a new race of malicious cyborgs, the Cybermen (Fig. 5.2), featured the first "regeneration" of the Doctor—passing the baton from actor William Hartnell to his successor Patrick Troughton, and succeeded in stemming the declining viewing figures of the series.[80]

In the early 1960s, developments in medical technologies, in particular organ transplants and "spare-part surgery", were receiving lots of press coverage, and commentators began to speculate about what the result of implanting machines into humans would be, and what such biological

[77] Jeremy Bentham, "Innes Lloyd Interview", *Doctor Who Magazine*, Winter Special, 11–13; and Chapman, 'Fifty Years in the TARDIS', 48–50; Michael Seely, *The Quest for Pedler: The Life and Ideas of Dr Kit Pedler* (Miwk Publishing, 2014), 65–69. For more on the differing importance of accuracy and plausibility in science fiction, and the authenticity and authority that scientific advisors can bring to a project see: Kirby, *Lab Coats in Hollywood*.

[78] *Tomorrow's World*, presented by Raymond Baxter, December 9, 1965, BBC One; Seely, *The Quest for Pedler*, 64.

[79] Seely, 69–75.

[80] The final episode of season four, serial one "The Smugglers" drew an audience of 4.4 million, "The Tenth Planet" incrementally built on this figure, with the episodes three and four achieving respectively 7.6 and 7.5 million viewers.

Fig. 5.2 The "Man-Amplifier" exoskeleton (left) created by Neil Mizen and his team at Cornell Aeronautical Lab, as featured in *New Scientist* in October 1965, closely resembled Pedler's Cybermen (right) as featured in "The Tenth Planet" serial. *Popular Science Monthly*, November 1965, 71; and BBC © BBC Photo Library

technology meant for the future of humanity.[81] With the Cybermen, Pedler, whose own research aimed to produce a "circuit diagram" of the retina, capitalised on these fears. Creating ruthless cyborgs who, while at their core were human-like beings from Earth's twin planet, had augmented themselves with machinery and now set about the universe changing other organic entities into Cybermen.[82] Over the coming years, as the writing partnership between Pedler and Davis flourished, this focus on exploring and exploiting society's techno-scientific anxieties, in particular those centred on human interference with nature and biological processes, would dominate the fictional worlds and narratives that the duo created.

In addition to the imaginative representation of new scientific ideas, the time-travelling element of *Doctor Who* also threw up many science-related philosophical and ethical considerations, such as the causal loop, or "bootstrap paradox" featured in season 20 serial "Terminus".[83] Other considerations on the philosophy of time travel featured in "Day of the Daleks" (1972), "City of Death" (1979) and "Genesis of the Daleks" (1975); the latter featuring a nod to time-travel ethics via a version of the "If I kill Hitler, could things be even worse?" trope.[84] In this serial, in which 12 years after their introduction, writer Terry Nation reimagined the Daleks' origin story, evolutionary themes on mutation and human-non-human boundaries abounded. As, Davros, the Chief Scientist of the humanoid Kaleds—depicted as Nazis, who are suffering a progressive

[81] See: John Davy, "Ways of improving 'spare-part surgery'", *The Observer*, July 7, 1963; Micahel Woodruff, "Surgical transplantation in theory and practice", *New Scientist*, October 3, 1963, 12–13; Alex Comfort, "Modified Men", *New Scientist*, August 19, 1965, 456–7.

[82] Pedler saw communication as a central part of scientists' role and regularly featured in popular media outlets discussing the specifics of his own research, reviewing popular science books, and musing on more general themes about the potential of medical research to improve the human condition. See: Cyril Dunn, "Doctor who creates Who" *The Observer*, February 5, 1967, 23; Kit Pedler, "Views of Perception", *The Guardian*, September 20, 1968, 8; Timothy Johnson, "Exploring the Jungle of the Eye", *London Illustrated News*, October 8, 1966, 12–13; and Kit Pedler, "Babies," (Letters) *Daily Mail*, March 3, 1970, 4.

[83] A causal loop is a theoretical idea that if someone could travel back in time they may interact in a series of events which ultimately leads to them in the future, causing a closed loop where the origin of the sequence of actions cannot be determined. Sometimes referred to as the bootstrap paradox in reference to the time-travel novella *By His Bootstraps* by Robert A. Heinlein first published in 1941. Jonathan Holmes, "Doctor Who: what is the Bootstrap Paradox?" *Radio Times*, October 10, 2015.

[84] For more on this trope and it use in science fiction see, "Hitler's Time Travel Exemption Act", TV Tropes Wiki, https://tvtropes.org/pmwiki/pmwiki.php/Main/HitlersTimeTravelExemptionAct *(last accessed 20/03/21).*

mutation as the result of a 1000-year nuclear war—attempted to reach their "final mutational form" creating the Daleks in the process.[85] How mutations could reach an identifiable endpoint, and why the Kaleds were attempting to get there quicker was never discussed.

The Doctor often attempted to stop evil enemies from adversely affecting the course of human history, and indeed sometimes tried to control and direct evolution itself. The issue of interfering in evolution was most directly addressed in the serial "Ghost Light" (1989) from the final season of the original *Doctor Who* run, which focused on religious contestation of evolution in late nineteenth-century England. However, despite repeated references to Darwin and evolution, the serial largely focused on the use of biological concepts to justify hierarchical structures, and was thus closer to the social Darwinian ideas of Darwin's contemporary Herbert Spencer (1820–1903).[86]

Despite its many seasons and multiple iterations, English Literature scholar, David Layton, identifies a philosophical consistency to *Doctor Who* that goes beyond the format itself. Layton argues that the show presents a "secular humanist view of the universe and humanity's place within it".[87] How consistent this secular humanist view really was, is debatable. Given its longevity and the history of guest writers contributing serials, over the seasons the series has had plotlines ranging from the outright mystical to the extreme secular rationalist.[88] While the secular humanist framing in *Doctor Who* may have been inconsistent, that it featured regularly is unsurprising given the wider culture

[85] Courtland Lewis and Paula J. Smithka, *Doctor Who and Philosophy: Bigger on the Inside* (Open Court Publishing, 2010), 180–183.

[86] Jonathan Dennis, "Scenes from the class struggle in Gabriel Chase: Darwinism, Social Darwinism, and Religion" in *Ghost Light*, The Black Archive, 6 (Obverse Books, 2016), 70–99. The effect of time traveller's actions on the course of history continues to be a regular theme of SFTV, for a recent example see *The Umbrella Academy* (Netflix, 2019–2020) based on the comic book series of the same name. For a literary account that explores the concept of divergent evolutionary pathways, see Kurt Vonnegut's *Galápagos* (1985), which explores the future evolution of humans after a global catastrophe leaves only an isolated group of fertile humans on the fictional Galápagos island of Santa Rosalia.

[87] David Layton, *The Humanism of Doctor Who: A Critical Study in Science Fiction and Philosophy* (McFarland, 2014), 2.

[88] For mystical serials, see "Kinda" (1982) and "Snakedance" (1983). While in several serials the Doctors' decisions are couched in terms of rational decision-making, perhaps the most clearly secular rationalist plotline can be found in the pilot for a *Dr Who* spin-off *K-9 and Company* (1981), in which science and technology are portrayed as good, and nature and ritual as bad.

of scientific programming at the BBC (Chaps. 2 and 3). That secular humanist themes were so central to such a long-running drama series alerts us to how, with perhaps the exception of the religion department, a secular humanist worldview came to dominate production culture across all BBC departments in the post-war decades.

From its mixed start in the 1960s, *Doctor Who* grew in success and became a flagship show of the BBC's weekend schedule, averaging 11.5 million viewers by the mid-1970s.[89] Following trends during the early 1970s in big-budget science documentaries (Chap. 6), the BBC increasingly syndicated and sold *Doctor Who* around the world.[90] As such, along with BBC natural history and science documentary output, *Doctor Who* should be considered part of the canon of British television shows that helped to globalise a distinctly British mode of presenting science.

Kit Pedler went on to contribute directly to three more *Doctor Who* serials, all featuring the Cybermen.[91] The BBC regularly promoted his scientific credentials, particularly when commentators called into question aspects of the show, such as the violent content in the final episode of his last serial "The Tomb of the Cybermen".[92] Pedler himself was comfortable in promoting the unusual dual-track of his career, regularly appearing in the popular press, radio and TV non-fiction features, and chat shows. It was on such a show, *My Life in Science*, in June 1969 when discussing his interest in the social effects of new scientific discoveries, that Pedler announced his latest science fiction creation, *Doomwatch* (1970–1972).[93]

[89] Chapman, 'Fifty Years in the TARDIS', 50.

[90] Beginning with New Zealand and Australia in 1964 and 1965 respectively, by the mid-1970s the series was shown on PBS stations in the US and Canada, and to date has regularly aired in an estimated 50 countries. In 2013, a 50th anniversary special set a world record, when it was simulcast and watched in 94 countries. Robert Booth, "Doctor Who one of biggest shows in the world, says BBC following 'simulcast'", *The Guardian*, November 24, 2013.

[91] "The Moonbase", *Doctor Who*, written by Kit Pedler and Gerry Davis (uncredited), February to March 1967, BBC One; "The Tomb of the Cybermen", *Doctor Who*, written by Kit Pedler and Gerry Davis, September 1967, BBC One; and "The Wheel in Space", *Doctor Who*, story by Kit Pedler, written by David Whitaker, April to June 1968, BBC One.

[92] Pedler appeared on the BBC panel debate show *Talkback* in September 1967 to defend the more violent aspects of the serial. The BBC then used Pedler's commentary, along with that of his co-panellist the social psychologist Dr Hilde Himmelweit to produce a report, which convinced Australian broadcaster ABC not to drop the last episode of the serial when they aired it in the summer of 1968. Seely, *The Quest for Pedler*, 114–119.

[93] "Of Ombudsmen and Cybermats", *My Life in Science*, presented by Elizabeth Gard and David Wilson, June 5, 1969, BBC Radio Three.

In the series, which aired in February 1970, Pedler again wrote in partnership with Gerry Davis (Fig. 5.3). Together, they swapped the fantastical planets and aliens of *Doctor Who* for a secretive government agency, the Department for the Observation and Measurement of Scientific Work, and their eponymous "analogue digital hybrid" computer, *Doomwatch*.[94] The agency, headed by Dr Quist, a Nobel prize-winning mathematician haunted by the guilt of his role on the Manhattan Project, was tasked with monitoring errant experimentation and new scientific applications in other branches of the government and the corporate world. Unorthodox but morally sound, Pedler and Davis cast Quist in the mould of Professor Quatermass, and similarly the series tapped into societal techno-anxieties of the early 1970s. The series was a big success, in part due to Davis and Pedler's ability to create storylines that were at the same time fantastical and scientifically grounded, resulting in arresting narratives that horrified audiences, in large part due to their scientific plausibility.[95] From the virus in episode one, created to decompose vast amounts of plastic waste onwards, many of the series' central themes proved incredibly prescient. However, Pedler and Davis' formula was straightforward. As with their creation of the Cybermen (Fig. 5.2), they simply gave the popular science press, in particular the *New Scientist,* a close reading, spinning off ideas from new scientific research and technological innovations.[96]

As with *Doctor Who*, on first viewing *Doomwatch* was not an obvious evolutionary fiction. However, with storylines in the first season alone covering topics including genetic engineering and the ethics of transplants (Ep 2, "Friday's Child"), selective breeding and eugenics (Ep 4, "Tomorrow, the Rat"), and chemical interference in sexual reproduction (Ep 11, "The Battery People"), I argue that *Doomwatch* should in fact be considered the archetypal evolutionary SFTV of the 1970s. To date, the series has more commonly been analysed through the lenses of the

[94] In the opening episode, the scientist Dr John Ridge (Simon Oates) introduced the new team member, Dr Toby Wren (Robert Powell) to the department's analogue digital hybrid computer, known as *Doomwatch*. However, as the series proceeds it becomes clear that *Doomwatch* is a nickname given to the government agency as a whole. *Doomwatch*, "The Plastic Eaters." Series 1: Episode 1. Produced by Terence Dudley. Written by Kit Pedler and Gerry Davis. BBC 1, February 9, 1970 (Simply Media/BBC – DVD, 2015): 03:50-04:10.

[95] Mark Wilson, 'Doomwatch and the Environment in Britain, 1970–c.1974', *Revue Française de Civilisation Britannique. French Journal of British Studies* 23, no. XXIII–3 (2018), https://doi.org/10.4000/rfcb.2621.

[96] Seely, *The Quest for Pedler*, 145–150.

Fig. 5.3 The *Doomwatch* writing team: (L-R) Gerry Davis, producer Terence Dudley, and Kit Pedler. © *Radio Times*

anti-nuclear and nascent environmental movements in the UK.[97] Indeed, Pedler was part of an early wave of scientists who were concerned with the environmental impacts of industrialised society, and as such *Doomwatch* presented some of the first dramatised television on pressing environmental matters (e.g. "Train and De-Train", 1970).[98] However, if we analyse

[97] Wilson, 'Doomwatch and the Environment in Britain, 1970–c.1974'.

[98] Environmental issues became increasingly important to Pedler throughout the 1970s, and by the end of the decade he had altered his lifestyle significantly in an attempt to cause

the three seasons as a whole, the most common plotlines are those centred more generally on issues arising from human interference with nature.[99] While this included many episodes with an environmental focus, it also aligned with Pedler's growing and increasingly public disdain for amoral and under-regulated scientific practices.[100] Thus, *Doomwatch* reflected, and in turn further popularised, wider societal anxieties on the role of science in society during the 1970s.

Counterculture, Doom and Declensionism

By the late 1960s, counter-cultural movements in both the USA and Europe were increasingly critiquing science's role in society.[101] Scientists and other public intellectuals were central to many of these movements.[102] In April 1969, several of the left-leaning scientists who had been involved with various science and society movements since the 1930s, including

the least disturbance and harm to other life forms, and was an advocate of James Lovelock's *Gaia Hypothesis*. See Kit Pedler, *The Quest for Gaia: A Book of Changes*, (Souvenir Press, 1979).

[99] The majority of episodes—10/13 in season one, 10/13 in season two and 9/12 in season three—have either a primary or secondary plotline focused on some aspect of human interference with natural processes or the natural environment. It should be noted that, after being marginalised in the process of writing and producing season two, Kit Pedler and Gerry Davis did not contribute to season three of *Doomwatch*, leaving the show after disagreements mounted with producer, Terence Dudley about the centrality of science to the stories and a shift to a more conservative dramatic focus for the series. See Seely, *The Quest for Pedler*, 184–191.

[100] Pedler often wrote letters to newspapers commenting on uncritical coverage of scientific issues. For examples, see a letter on unnecessary animal experimentation by scientists (Kit Pedler, "Letters: These cruel scientists go too far", *Daily Mail*, August 24, 1970, 2), and a letter on his concerns with experiments developing test-tube babies (Kit Pedler, "Babies", *Daily Mail*, March 3, 1970, 4).

[101] Simone Turchetti, 'Looking for the Bad Teachers: The Radical Science Movement and Its Transnational History', in *Science Studies during the Cold War and Beyond: Paradigms Defected*, ed. Elena Aronova and Simone Turchetti, Palgrave Studies in the History of Science and Technology (Palgrave Macmillan US, 2016), 77–101, https://doi.org/10.1057/978-1-137-55943-2_4.

[102] For example, the Campaign for Nuclear Disarmament, founded in 1957, who's first President was the philosopher Bertrand Russell, the physicist Joseph Rotblat was on the initial executive committee, and biologists Julian Huxley and Conrad Waddington were among a large number of public intellectuals who supported the organisation in its early years. Christopher P. Driver, *The Disarmers: A Study in Protest* (Hodder and Stoughton, 1964), 42–45.

Lancelot Hogben and Julian Huxley, gave their support to a new organisation, the British Society for the Social Responsibility in Science (BSSRS).[103] With roots in the wider counter-cultural movements of 1960s London, the initial aims of the BSSRS were to increase scientists' awareness of their social responsibility, to draw attention to the social pressures influencing science, and to inform the public about the implications of scientific developments.[104] The BSSRS had grown out of campaigns against biological weapons, and one of its first meetings led to the publication of *The Social Impact of Modern Biology*, which featured several well-known scientists, including the French biochemist Jacques Monod (1910–1976) and the mathematician and broadcaster Jacob Bronowski, who we first met in Chap. 3 (for more on both see Chap. 6).[105] The BSSRS grew quickly, and by the early 1970s was part of a burgeoning "radical science" movement in the UK.[106] In 1971, as part of a round of publicity promoting the need for a real-life *Doomwatch* department, which featured in the *New Scientist*, the BSSRS invited Pedler to speak at their Imperial College branch in London.[107] In the early 1970s, Pedler also advised, supported and contributed to the radical science and alternative technology magazine *Undercurrents*.[108] However, the combination of his popular fiction writing, the time his non-academic roles were now taking, and the increasingly outspoken and public positions he was taking on many aspects of institutionalised science had already begun to create tensions with his colleagues at the Institute of Ophthalmology. After handing in his Home Office license for animal experimentation in 1970, in the

[103] Jon Agar, 'What Happened in the Sixties?', *The British Journal for the History of Science* 41:4 (2008): 567–600; David Kaye (1986), *The Life and Times of BSSRS*, http://www.bssrs.org/services *(last accessed 14/06/21).*

[104] Kaye, *The Life and Times of BSSRS*, 11–12 and Appendix 5, "BSSRS Constitution"

[105] Fuller Watson, *The Social Impact of Modern Biology* (Routledge, 1971).

[106] See Jonathan Rosenhead, "The BSSRS: three years on", *New Scientist*, April 20, 1972, 134–136.

[107] Peter Elphick, "Social Responsibility in Science: New I.C. Society", *Felix: Imperial College Union*, 2. Available online: https://f001.backblazeb2.com/file/felixonlinearchive/issues/pdfs/felix_303.pdf (last accessed 18/09/2020); and Graham Chedd, "Doomwatcher incarnate", *New Scientist and Science Journal*, March 18, 1971, 622–624.

[108] Pedler was regularly listed among those who had supported and advised the magazine in editions from its launch in 1972–1975, and in 1976 he contributed the article: Kit Pedler, "Peoples Habitat: Special Feature", *Undercurrent*, 16 (June-Jul 1976), 23, https://undercurrents1972.wordpress.com/2013/02/09/uc16-june-july-1976/ *(last accessed 21/06/2021).*

summer of 1971 Pedler made the decision to resign from his role as the Director of the Anatomy Department.[109]

In the storylines of series one of *Doomwatch*, Pedler and Davis successfully managed the balance between scientific critique and plausibility. More than simply reflecting societal concerns, *Doomwatch* was bound up in there actual creation and popularisation. As one rather sensationalist piece in the *Daily Mail* reflected, "this science fiction series has an uncanny knack of turning into fact", or as Gerry Davis put it, "[w]e feel we are putting out a public service warning people what could—and does—happen when progress marches forward too fast".[110] As this quote alludes to, while plotlines about emergent biological technologies and the bioethics of their potential uses abounded, *Doomwatch* most certainly did not reproduce the progressive narrative arc of the evolutionary epic. Rather, the series co-opted the declensionist narrative of the nascent environmental movement—a stark warning about the limitations and unexpected consequences of techno-scientific solutions. Factual BBC programming from the period also reflected these same concerns, most notably a BBC Community Programme Unit-produced show made by the BSSRS, *Open Door* (1974), and the science debate show, *Controversy* (1971–1975) (Chap. 6). However, as we shall see in more depth in Chap. 6, the most successful mainstream science documentaries of the 1970s remained wedded to an inherently progressive, positivist vision of science.

The framing of science in *Doomwatch* largely centred on conflict and controversy. To assist audience comprehension, Pedler and Davis regularly foregrounded contentious new scientific ideas and technologies by referencing historical events and discussions that would have been familiar to the general population. For example, in "Tomorrow, the Rat" (S1 Ep 4), the script made repeated references to Nazi eugenics programmes: "they're like embryonic Nazis on four legs!"[111] This episode painted the errant geneticist Dr Mary Bryant's experiments as immoral and contrasted them repeatedly against the natural order and "God's law".[112] Although some depth was added to the character when the audience learned her

[109] Seely, *The Quest for Pedler*, 212–215; Chedd, "Doomwatcher incarnate", 622–624; and Raymond Gardner, "Prophets of Doom", *The Guardian*, December 13, 1973, 11.

[110] Daily Mail Reporter, "TV's crystal ball one step ahead of reality", *Daily Mail*, March 13, 1970, 12.

[111] *Doomwatch*, "Tomorrow, the Rat". Series 1: Episode 4. Written and produced by Terence Dudley. BBC 1, March 2, 1970 (Simply Media/BBC – DVD, 2015): 39:50-39:55. All BBC copyright content reproduced courtesy of the British Broadcasting Corporation. All rights reserved.

[112] *Ibid*: 21:40.

motivation was to reduce human suffering, Dr Bryant embodied many of the traits of the mad-scientist trope. In a common science fiction narrative, familiar since Mary Shelley's gothic classic *Frankenstein* (1818), in the closing sequence Dr Bryant's hubris was confirmed, as the genetically modified rats, that she had dedicated her life to creating, killed her. Although Pedler and Davis' input and influence on the series declined from season two onwards, the trope of the morally reprehensible scientist persisted. For example, in season three's "The Hair Trigger", mental patients were treated by computers to speed up the process of natural selection, in socially and ethically contentious scientific experiments.[113] Again, the episode alluded to the more questionable past applications of evolutionary theory, such as eugenics, and drew upon the longstanding cultural trope of the morally reprehensible mad scientist who interferes with natural processes.[114] Here, as throughout the series, while science or more specifically, scientific ethics were called into question, the veracity of evolutionary theory itself was never in doubt.

The moral and ethical questions interrogated in these two episodes of *Doomwatch* mirrored the debate in 1958 on the final episode of the science series *Five Hundred Million Years* (Chap. 3). In *Five Hundred Million Years* the inclusion of religious figures explicitly framed the discussion on the relationship between evolution and religion. Over a decade later, within the fictional world of *Doomwatch*, the narrative was shorn of its supernatural elements and reframed as a discussion about societal norms and the ethics of scientific practice. However, as the repeated references to God in *Doomwatch* highlight—"with a genetically perfect reproduction you may come face to face with God"[115]—it is not easy to disentangle questions such as whether a man has the right to interfere in his own biological development, from their religious origins.[116] Further, one of the

[113] *Doomwatch*. "The Hair Trigger." Series 3: Episode 6. Directed by Quentin Lawrence. Written by Brian Hayles. BBC 1, July 10, 1972 (Simply Media/BBC – DVD, 2015).

[114] For more on eugenics in science fiction see David A. Kirby, "The Devil in Our DNA: A Brief History of Eugenics in Science Fiction Films," *Literature and Medicine* 26:1 (2007): 83–108, https://doi.org/10.1353/lm.2008.0006. For more on mad-scientist narratives and tropes, see Spencer R. Weart, *The Rise of Nuclear Fear* (Harvard University Press, 2012), 32–54.

[115] *Doomwatch*, "Tomorrow, the Rat". Series 1: Episode 4. Written and produced by Terence Dudley. BBC 1, March 2, 1970 (Simply Media/BBC – DVD, 2015): 36:50.

[116] For two examples from many, which explore in-depth some of the intertwined histories of ethical and religious contestations of evolution, see: Peter J. Bowler, *Evolution: The History*

effects of SFTV in this tradition is to reinforce the stereotypical association between evolutionary science and controversy, whether ethical, moral or religious. During the second half of the twentieth century, the idea that scientific research in this area was controversial and contentious on moral and ethical grounds was utilised by minority-organised religious groups attempting to undermine the authority of the biological sciences (see Chap. 7). In particular, in the USA in an attempt to legitimise their beliefs, the Intelligent Design movement relegated traditional scriptural or theological objections for pseudo-scientific and ethical arguments against evolutionary theory.[117]

Science fiction is not the only area of broadcast media that we find scientists and scientific practice reduced to simplistic caricature.[118] By the early 1970s, the BBC accompanied any relatively large-budget series, such as *Doomwatch*, with a number of co-ordinated promotional activities. As we have seen with the example of Kit Pedler included in this chapter, it was often in these more reflective environments when writers and producers discussed the broader motivations behind a series. By the 1970s, science broadcasters and popular scientists were also undertaking the same promotional circuit to endorse their own works of non-fiction, such as popular science books and documentary series. As we will see in Chap. 7, on occasion this promotional media enabled popular scientists with fringe or pseudo-scientific theories on evolution to bypass the structures and gatekeepers of more formal scientific broadcast content.

The consideration of scientific futures and humanity's evolutionary future has not been solely the preserve of SFTV, and sometimes the boundary between fiction and non-fiction has been almost indistinguishable. Take *Tomorrow's World* (1965–2003) a long-running BBC science and technology magazine show usually concerned with reviewing the latest technological innovations and gadgets. On their 500th episode in 1978, producers flipped the format to "fantasise a little about what future

of an Idea (University of California Press, 1989); and Gowan Dawson, *Darwin, Literature and Victorian Respectability* (Cambridge University Press, 2007).

117 For more on the history of the Intelligent Design movement in the USA see: Tom Kaden, *Creationism and Anti-Creationism in the United States: A Sociology of Conflict* (Springer, 2018).

118 For more on the most common representations of scientists in fiction see: Petra Pansegrau, "Stereotypes and Images of Scientists in Fiction Films" in Peter Weingart and Bernd Huppauf (Eds), *Science Images and Popular Images of the Sciences* (Routledge, 2012), 257–266.

Tomorrow's Worlds might bring to our children's children".[119] Rather than making rigorous predictions about what the future actually held for the British public, like *Doomwatch,* this episode of *Tomorrow's World* more accurately reflected societal concerns about environmental degradation at the time. This blurred boundary is considered further in the following chapter, in particular in the most ambitious show on evolution of the 1970s, *The Life Game* (1973).

Conclusion

In analysing fictional content with evolutionary themes, and the cultural contexts in which shows emerged, I want to encourage you to reflect on the potential cumulative effect that media consumption can have on influencing individual viewers' understanding of, and affinity towards, specific scientific subjects. While clearly varying in specifics across genres—particularly the fiction/non-fiction divide—there are many tropes, narrative devices and visual elements common across BBC output that features evolutionary science. Unlike non-fiction science programming, *Doctor Who* provided a space where evolutionary ethics could be playfully interrogated, allowing writers to play around with the timeline of the evolutionary epic, and more directly comment on its central progressive narrative. Thus, while British SFTV may not have directly replicated the narrative of the evolutionary epic, instead more regularly featuring cautionary declensionist plotlines, usually it still relied on scientific progressivism and positivism, exemplified by techno-scientific fixes and the morally principled scientists who came up with them. Further, in keeping with the dominant secular humanist worldview at the centre of BBC science programming during the twentieth century, across all of these genres the natural theological language of awe, wonder and spectacle, was repurposed for non-theistic contexts (see Chap. 6).

Many popular fictional depictions of evolution on radio and television reflected and reinforced popular (mis)understandings of evolutionary theory already held by sections of the British public. While those producing

[119] "500th Programme", *Tomorrow's World*, March 16, 1978, BBC One, 00:40-00:50. As originally featured on the BBC Archive online in 2018, see: https://web.archive.org/web/20190529173410/http://www.bbc.co.uk:80/archive/tomorrowsworld/. For more clips of *Tomorrow's World* from the late 1960s to the 1980s, see: https://www.bbc.co.uk/archive/tomorrows-world/zrkpwty *(last accessed 21/06/2021)*.

the shows were merely utilising standard modes of plot and narrative development, the controversial and contested depiction of science on which they so often relied inadvertently further reinforced a common historical narrative, which placed evolutionary theory in conflict with the expected moral or ethical norms of society. These shows presented a valid interrogation of the issues, which served a useful societal function. However, they also remind us that as with the example of the Intelligent Design movement, different interest groups across society will always understand and utilise these associations in divergent ways.

In briefly outlining some of the diverse and often hard to locate instances when evolution featured on non-science BBC content, I have introduced the idea that although diffuse, these shows often presented the most stereotypical framing of evolution. The perpetuation of these stereotypical tropes is most notable on television drama content that featured evolution, whether the ethically questionable science fiction of *Doomwatch* or in the many historical dramatisations centred on Darwinian controversies. Unlike in educational broadcasting, where programmes often featured the latest cutting-edge developments in evolutionary biology, in most of the non-science content, depiction of evolutionary science was often outdated, incomplete, teleological and, on occasion, simply incorrect. Although coverage of evolution in these shows was often fleeting, the pernicious association of the subject with morally ambiguous or reprehensible practices across a range of broadcast media, movies and science fiction novels from the twentieth century has created a normative frame that explicitly links evolutionary ideas to controversy and conflict. At the societal level, the cumulative effect of the repeated use of this media frame is likely to reinforce the stereotypical social norms that associate evolution and religion with conflict.[120] While the direct effects of consuming any one piece of media are almost impossible to discern, when we begin to consider an individual's media consumption as an often self-selecting, cumulative process, this ramshackle collection of chat shows, fiction and dramas that feature across the corpus take on an increased importance.

[120] For an introduction to the concept of media reinforcement and cumulative effects of media consumption see: Michael D. Slater, 'Reinforcing Spirals: The Mutual Influence of Media Selectivity and Media Effects and Their Impact on Individual Behavior and Social Identity', *Communication Theory* 17:3 (2007): 281–303, https://doi.org/10.1111/j.1468-2885.2007.00296.x.

References

Agar, Jon. 'What Happened in the Sixties?' *The British Journal for the History of Science* 41:4 (2008): 567–600.

Avery, Todd. *Radio Modernism: Literature, Ethics, and the BBC, 1922–1938*. Ashgate Publishing, Ltd., 2006.

Barnett, Richard. 'Education or Degeneration: E. Ray Lankester, H. G. Wells and The Outline of History'. *Studies in History and Philosophy of Science Part C: Studies in History and Philosophy of Biological and Biomedical Sciences* 37:2 (2006): 203–29. https://doi.org/10.1016/j.shpsc.2006.03.002.

Batchelor, John. *H. G. Wells*. Cambridge University Press, 1985.

BBC. *BBC Year-Book* 1934. BBC, 1934.

———. BBC Year-Book 1947. BBC, 1947.

Bentham, Jeremy. "Innes Lloyd Interview", *Doctor Who Magazine*, Winter Special, 11–13.

Blancke, Stefaan. 'Lord Monboddo's Ourang-Outang and the Origin and Progress of Language'. In *The Evolution of Social Communication in Primates: A Multidisciplinary Approach*, edited by Marco Pina and Nathalie Gontier, 31–44. Interdisciplinary Evolution Research. Springer International Publishing, 2014. https://doi.org/10.1007/978-3-319-02669-5_2.

Booth, Robert. "Doctor Who one of biggest shows in the world, says BBC following 'simulcast'", *The Guardian*, November 24, 2013.

Bould, Mark, and Sherryl Vint. *The Routledge Concise History of Science Fiction*. Routledge, 2011.

Bowler, Peter J. *Evolution: The History of an Idea*, University of California Press, 1989.

———. A History of the Future: Prophets of Progress from H. G. Wells to Isaac Asimov. Cambridge University Press, 2017.

———.'Parallel Prophecies: Science Fiction and Futurology in the Twentieth Century'. *Osiris* 34:1 (2019): 121–38. https://doi.org/10.1086/703952.

Brooke, John Hedley, John Brooke, and Geoffrey Cantor, *Reconstructing Nature: The Engagement Of Science And Religion*. A&C Black, 2000.

Buckell, Gareth J. *Rayner Heppenstall: A Critical Study*. Dalkey Archive Press, 2007.

Bud, Robert. '"The Spark Gap Is Mightier than the Pen": The Promotion of an Ideology of Science in the Early 1930s'. *Journal of Political Ideologies* 22:2 (2017): 169–81. https://doi.org/10.1080/13569317.2017.1298548.

Chapman, James. 'Fifty Years in the TARDIS: The Historical Moments of Doctor Who'. *Critical Studies in Television* 9:1 (2014): 43–61. https://doi.org/10.7227/CST.9.1.4.

Chedd, Graham. "Doomwatcher incarnate", *New Scientist and Science Journal*, March 18, 1971, 622–624.

Comfort, Alex. "Modified Men", *New Scientist*, August 19, 1965, 456–7.
Daily Mail Reporter, "TV's crystal ball one step ahead of reality", *Daily Mail*, March 13, 1970, 12.
Davy, John. "Ways of improving 'spare-part surgery'", *The Observer*, July 7, 1963.
Dawson, Gowan. *Darwin, Literature and Victorian Respectability*. Cambridge University Press, 2007.
de la Torre, Lillian. *Dr. Sam: Johnson, Detector*. Alfred A. Knopf, 1946.
Dennis, Jonathan. *Ghost Light*. Edited by Philip Purser-Hallard. The Black Archive 6. Obverse Books, 2016.
Drakakis, John. *British Radio Drama*. Cambridge University Press, 1981.
Driver, Christopher P. *The Disarmers: A Study in Protest*. Hodder and Stoughton, 1964.
Duke, D. "Like Carroon", *Radio Times*, September 4, 1953, 9.
Dunn, Cyril. "Doctor who creates Who" *The Observer*, February 5, 1967, 23.
Elphick, Peter. "Social Responsibility in Science: New I.C. Society", *Felix: Imperial College Union*, 2. https://f001.backblazeb2.com/file/felixonlinearchive/issues/pdfs/felix_303.pdf
Elsdon-Baker, Fern, and Bernard Lightman. *Identity in a Secular Age: Science, Religion, and Public Perception*. University of Pittsburgh Press, 2020.
Frattarola, Angela. 'The Modernist "Microphone Play": Listening in the Dark to the BBC'. *Modern Drama* (2010), https://doi.org/10.3138/md.52.4.449.
Gardner, Raymond. "Prophets of Doom", *The Guardian*, December 13, 1973, 11.
Goldsmith, Maurice. *The Science Critic: A Critical Analysis of the Popular Presentation of Science*. Routledge & Kegan Paul, 1986.
Harding, Gilbert. "Gilbert Harding's Notebook", *Picture Post*, November 14, 1953, 40.
Heppenstall, Rayner (Ed.), *Imaginary Conversations*. Secker & Warburg, 1948.
Heppenstall, Rayner. *Four Absentees*. Barrie and Rockliff, 1960.
Hochscherf, Tobias, James Leggott, and Donald E. Palumbo. *British Science Fiction Film and Television: Critical Essays*. McFarland, 2014.
Holmes, Jonathan. "Doctor Who: what is the Bootstrap Paradox?" *Radio Times*, October 10, 2015.
Johnson, Timothy. "Exploring the Jungle of the Eye", *London Illustrated News*, October 8, 1966, 12–13.
Johnston, Derek. "Experimental Moments: R.U.R. and the Birth of British Television Science Fiction". *Science Fiction Film & Television* 2:2 (2009): 251–68. https://doi.org/10.1353/sff.0.0079.
———. "Genre, Taste and the BBC: The Origins of British Television Science Fiction" (University of East Anglia, PhD Thesis, 2009).
Kaden, Tom. *Creationism and Anti-Creationism in the United States: A Sociology of Conflict*. Springer, 2018.
Kaye, David. *The Life and Times of BSSRS* (1986), http://www.bssrs.org/services.

Kirby, David A. 'Are We Not Men?: The Horror of Eugenics in The Island of Dr. Moreau'. *ParaDoxa* 17 (2002). https://www.research.manchester.ac.uk/portal/en/publications/are-we-not-men-the-horror-of-eugenics-in-the-island-of-dr-moreau(7559fa88-776f-4bbf-aa6a-5f16ad76ca1f)/export.html.

———. Lab Coats in Hollywood: Science, Scientists, and Cinema. MIT Press, 2011.

———. "The Devil in Our DNA: A Brief History of Eugenics in Science Fiction Films". *Literature and Medicine* 26:1 (2007): 83–108. https://doi.org/10.1353/lm.2008.0006.

Layton, David. *The Humanism of Doctor Who: A Critical Study in Science Fiction and Philosophy*. McFarland, 2014.

Lejeune, C. A. "Television Notes", *The Observer*, July 26, 1953, 11.

Lewis, Courtland, and Paula J. Smithka. *Doctor Who and Philosophy: Bigger on the Inside*. Open Court Publishing, 2010.

Lightman, Bernard. "Victorian Sciences and Religions: Discordant Harmonies," *Osiris*, Science in Theistic Contexts: Cognitive Dimensions, 16 (2001): 343–66.

McCoy, Glen. *Dr Who: Timelash*. W. H. Allen.

Mittell, Jason. 'Narrative Complexity in Contemporary American Television'. *The Velvet Light Trap* 58:1 (2006): 29–40. https://doi.org/10.1353/vlt.2006.0032.

Morris, Desmond. "Science and Religion: Exploring the Spectrum, Life Story Interviews: Desmond Morris," National Life Stories, C1672/16, The British Library, https://sounds.bl.uk/Oral-history/Science/021M-C1672X0016XX-0001V0.

Orthia, Lindy A. 'Antirationalist Critique or Fifth Column of Scientism? Challenges from Doctor Who to the Mad Scientist Trope', *Public Understanding of Science* 20:4 (2011).

Orwell, George. *Smothered Under Journalism 1946*. Secker & Warburg, 2001.

———."The Voyage of the 'Beagle'", *Radio Times*, March 22, 1946.

Pedler, Kit. "Views of Perception", *The Guardian*, September 20, 1968, 8.

———."Babies," (Letters) *Daily Mail*, March 3, 1970, 4.

———."Letters: These cruel scientists go too far", *Daily Mail*, August 24, 1970, 2.

———. "Peoples Habitat: Special Feature", *Undercurrent*, 16 (June-Jul 1976), 23.

———.The Quest for Gaia: A Book of Changes. Souvenir Press, 1979.

Radio Critic, "Saturday Serial", *The Manchester* Guardian, August 5, 1953, 3.

Rees, Amanda, and Charlotte Sleigh. *Human*. Reaktion Books, 2020.

Rieder, John. *Colonialism and the Emergence of Science Fiction*. Wesleyan University Press, 2012.

Rosenhead, Jonathan. "The BSSRS: three years on", *New Scientist*, April 20, 1972, 134–136.

Scannell, Paddy, and David Cardiff. *A Social History of British Broadcasting: Volume 1 – 1922–1939, Serving the Nation*. Wiley, 1991.

Schwartz, A. Brad. *Broadcast Hysteria: Orson Welles's War of the Worlds and the Art of Fake News*. Farrar, Straus and Giroux, 2015.
Sconce, Jeffrey. *Haunted Media: Electronic Presence from Telegraphy to Television*. Duke University Press, 2000.
Seely, Michael. *The Quest for Pedler: The Life and Ideas of Dr Kit Pedler*. Miwk Publishing, 2014.
Slater, Michael D. 'Reinforcing Spirals: The Mutual Influence of Media Selectivity and Media Effects and Their Impact on Individual Behavior and Social Identity', *Communication Theory* 17:3 (2007): 281–303, https://doi.org/10.1111/j.1468-2885.2007.00296.x.
Slusser, George Edgar, Patrick Parrinder, and Danièle Chatelain. *H.G. Wells Perennial Time Machine*. University of Georgia Press, 2001.
Smith, David C. *H. G. Wells, Desperately Mortal*. Yale University Press, 1988.
Swift, John. "To the World's End in Sixty Minutes", *Radio Times*, January 21, 1949, 25.
Telotte, J. P. *The Essential Science Fiction Television Reader*. University Press of Kentucky, 2008.
"The Vanishing Pulps", *Radio Times*, December 1, 1960, 29.
Thomas, P. "Thwarted" *Radio Times*, September 4, 1953, 9.
Thompson, Kristin. *Storytelling in Film and Television*. Harvard University Press, 2003.
Thumim, Janet. *Small Screens, Big Ideas: Television in the 1950s*. Bloomsbury Academic, 2002.
"Timelash" *Doctor Who in Vision*, Issue 83, March 1999.
Tulloch, John, and Manuel Alvarado. *Doctor Who: The Unfolding Text*. St. Martin's Press, 1983. http://archive.org/details/doctorwhounfoldi00tull.
Turchetti, Simone. 'Looking for the Bad Teachers: The Radical Science Movement and Its Transnational History'. In *Science Studies during the Cold War and Beyond: Paradigms Defected*, edited by Elena Aronova and Simone Turchetti, 77–101. Palgrave Studies in the History of Science and Technology. Palgrave Macmillan US, 2016. https://doi.org/10.1057/978-1-137-55943-2_4.
"Viewers are Saying", *Radio Times*, February 4, 1949, 25.
Watson, Fuller. *The Social Impact of Modern Biology*. Routledge, 2014.
Weart, Spencer R. *The Rise of Nuclear Fear*, Harvard University Press, 2012.
Weingart, Peter, and Bernd Huppauf. *Science Images and Popular Images of the Sciences*. Routledge, 2012.
Wilson, Mark. 'Doomwatch and the Environment in Britain, 1970–c.1974'. *Revue Française de Civilisation Britannique. French Journal of British Studies* 23, no. XXIII–3 (2018). https://doi.org/10.4000/rfcb.2621.
Woodruff, Michael. "Surgical transplantation in theory and practice", *New Scientist*, October 3, 1963, 12–13.

CHAPTER 6

Humanist Blockbusters: Depicting the Evolutionary Epic on Television

One basic idea that runs through all the sciences today is the idea of evolution, which was first thought of in biology. We now recognise that living things are not alone in having evolved from simple beginnings to their high and integrated complexity. Dead matter also evolves in time, step by step, from simple to complex; even the atoms of physics are all built up in the stars constantly from units of hydrogen. And at the other extreme, in the human mind, the gifts of reason and imagination are also built up from simple units, step by step from birth onwards.

There is an element of chance in every process of evolution. … It is another new and universal idea in science that order on the large scale is made from disorder on the small scale. And then in time, every large order will fall apart, by the erosion of chance, to its underlying dance of disorder. That is the very nature of time.[1]

In June 1969, Aubrey Singer, now Head of the BBC Features Group, and Robert Reid, who was Head of Science and Features, met with Jacob Bronowski in Paris to try to convince him to come on board for a major new history of science series.[2] The series was to follow the format of the

[1] Jacob Bronowski, *Insight* (Macdonald and Co., 1964), 7.

[2] Robert Reid was originally a science journalist and broadcaster who had worked under Singer when he was Head of Outside Broadcasts in the late 1950s (Chap. 3). In 1961, Singer had been appointed Head of Science and Features, and in 1968 his title was changed to Head

A. Hall, *Evolution on British Television and Radio*, Palgrave Studies in Science and Popular Culture,
https://doi.org/10.1007/978-3-030-83043-4_6

historian Kenneth Clark's documentary series *Civilisation* (1969), which told the history of man through art.[3] Commissioned by David Attenborough, who was now Controller of BBC2, Singer was surprised that a "scientific man" such as Attenborough had tackled art before science, and encouraged him to commission a follow-on that focused on the history of science.[4] Bronowski, who since 1964 had been a senior fellow at the Salk Institute for Biological Studies in San Diego, California,[5] was Singer's preferred choice to front the new "Story of Science" series.[6]

Since *Five Hundred Million Years* (1958), both the scale and scope of BBC science features had grown considerably. In 1960–1961 Bronowski had presented *Insight*, one of the first BBC television series dedicated to the development of science.[7] The series was an ambitious attempt to show that science was more than just facts and methods, that is was a "web of ideas" and "a vision of nature".[8] Despite BBC producer Gerald Leach

of Features Group. *The British Broadcasting Corporation Handbook* (HMSO, 1968), 172 & 182; Michael Leapman, "Singer, Aubrey Edward (1927–2007)", *Oxford Dictionary of National Biography*, Oxford University Press, 2011, http://www.oxforddnb.com/view/article/98830.

[3] The BBC's second television channel, BBC2 had launched in April 1964, and following the promotion of its first Controller Michael Peacock to Controller of BBC 1 in 1965, David Attenborough was appointed Controller of the channel, a position he would occupy until 1969. See Asa Briggs, *The History of Broadcasting in the United Kingdom: Volume V: Competition* (OUP, 1995), 402–415; Jean-Baptiste Gouyon, *BBC Wildlife Documentaries in the Age of Attenborough* (Springer Nature, 2019), 125–126.

[4] Marcus Hearn, "The Ascent of Man – Viewing Notes," BBC DVD/BBC Worldwide, 2005, 7; and David Attenborough, *Life on Air: Memoirs of a Broadcaster* (Princeton University Press, 2002), 214.

[5] Bronowski, along with the molecular biologist Francis Crick, and the French biochemist Jacques Monod, had been informally advising Jonas Salk, the institute's founder since before its official launch in December 1960. Francis Crick, *What Mad Pursuit* (Basic Books, 2008), 145. For more on the founding of the Salk Institute, and the influence of British humanist scientific intellectuals, particularly Bronowski and his contemporary C. P. Snow see: Robert Westman, 'The "Two-Cultures" Question and the Historiography of Science in the Early Decades of the Salk Institute for Biological Studies', *Sartoniana* 32 (2019), 43–85, http://www.sartonchair.ugent.be/file/293.

[6] Humphrey Fisher, "The Story of Science," BBC Memorandum, 31 January 1968, T64/64/1, BBC-WA.

[7] *Insight*. Produced by Humphrey Fisher and Bill Wright. Written by Jacob Bronowski, Gerald Leach and Bryan Silcock, and presented by Jacob Bronowski. BBC TV, November 1960–April 1961. For more on the content of *Insight* see the book that followed the series: Bronowski, *Insight*.

[8] Bronowski, 8.

Table 6.1 Episodes of *Insight* (1960–1961) written and presented by Jacob Bronowski, highlighting the theoretical focus and thematic structure of the series

Episode	*Title*	*Broadcast*
1	The Birth of Understanding	1 November 1960
2	The Birth of Matter	8 November 1960
3	The Shapes of Space	6 December 1960
4	The Shapes of Life	13 December 1960
5	Looking at Chance	3 January 1961
6	Chance or Choice	10 January 1961
7	Time is the Arrow	7 February 1961
8	Life and the Arrow of Time	14 February 1961
9	Language—The Tool of Ideas	7 March 1961
10	The Idea of Relativity	14 March 1961
11	The Vision of Our Age	4 April 1961

(1933–2004) and science reporter Bryan Silcock (1933–2019) providing research, visual aid development and scriptwriting support, Bronowski's ambitious thematic structure (Table 6.1) proved difficult to communicate from a studio setting.[9] Hampered by an intermittent broadcast schedule, and what one critic described as Bronowski's "wilful obscurantism", the series was not a success.[10] When in 1958–1959 the BBC General Advisory Council reviewed science broadcasting, the pioneer of management cybernetics, Stafford Beer (1926–2002) described *Insight* as typical of an "apologetic presentation", adding that its cabaret-style left him expecting "a flight of doves to emerge from the person of Dr Bronowski".[11] While not a success, *Insight* did contain several elements deemed worthy of future development. When Bronowski's didactic demonstrations and pre-shot

[9] Both Leach and Silcock would go on to have successful careers as science journalists, Silcock worked for *The Times* for 34 years, while Leach wrote for *The* Guardian, *The Sunday* Times, and *The Observer* and is remembered for his award-winning articles on the impact of science on society. Pearce Wright, "Gerald Leach", *The Guardian Online*, January 21, 2005, https://www.theguardian.com/science/2005/jan/21/obituaries.pressandpublishing *(last accessed 01/10/2020)*; and "Bryan Silcock Obituary", *The Times Online*, January 24, 2020, https://www.thetimes.co.uk/article/bryan-silcock-obituary-5ddhpz6vl *(last accessed 01/10/2020)*.

[10] Mary Crozier, "Television", *The Guardian*, January 4, 1961, 5; Maurice Richardson, "Critic's Narrow Escape", *The Observer*, February 12, 1961, 31.

[11] Stafford Beer, "Minutes of the General Advisory Council", 1959, 24, T16/623, BBC-WA; BBC, *Annual Report and Accounts for the year 1958–1959* (HMSO, 1959), 12.

illustrations dovetailed with relatable societal problems, executives saw the potential of his encompassing thematic approach.[12] As one TV critic summarised, "Science TV is on a good wicket with this kind of stuff. It can hold a peak-hour timing when it discusses the world that most people are curious about."[13] Indeed, during discussions for the new history of science series in 1969, Singer reflected that *Insight* was at least five years ahead of its time.[14]

Despite its thematic structure and the lack of a clear narrative arc, *Insight* was not without a central thesis. As the quote opening this chapter illustrates, although it lacked a clear chronological structure—there was no overt evolutionary epic—the series was situated within a cosmic evolutionary framework. Bronowski drew connections across the sciences to explain the observation of increasing complexity across time. In the final episode, in a glimpse of what was to come almost a decade later, Bronowski left the studio, discussing the history of science and art from atop of the Acropolis in Athens.[15]

The rest of this chapter outlines how following Aubrey Singer's approach in 1969, Bronowski's ambition to create a singular narrative on science, finally found fertile ground at the BBC, coinciding with an expansive period in television documentary production. The result *The Ascent of Man* (1973) was a series that helped to codify a televisual format that was capable of presenting a critical, but ultimately unswerving narrative of scientific progress. Built on the foundations of Huxley's evolutionary humanism, this is a sub-genre of science documentary that I call the humanist blockbuster.[16] Bronowski's series was somewhat at odds with other

[12] For a good example of when this worked well in the series, see episode 4 "The Shapes of Life", in which Bronowski combined a studio model illustrating mutations in a chromosome with the latest microscope images of human chromosomes, and the insights of geneticist Professor Lionel Penrose, who joined Bronowski in the studio to discuss his research on hereditary. "The Shapes of Life", *Insight*, episode 4, written by Jacob Bronowski and Gerald Leach, presented by Jacob Bronowski, BBC Television, December 13, 1960—transcript available at BBC-WA.

[13] Peter Black, "Peter Black's Teleview", *Daily Mail*, December 14, 1960, 3.

[14] Correspondence from Aubrey Singer to Jacob Bronowski, 15 August 1969, T64/64/1, BBC-WA.

[15] Mary Crozier, "Television", *The Guardian*, February 15, 1961, 7; and Mary Crozier, "Television", *The Guardian*, April 5, 1961, 5.

[16] For more on the sub-genre of humanist blockbusters, and in particular the history of the BBC religion department in blocking the broadcasting of humanist worldviews, see: Alexander Hall, 'A Humanist Blockbuster? Jacob Bronowski and The Ascent of Man', in

approaches to popular science during the early 1970s, which were being influenced by a range of counter-cultural movements. After detailing how Bronowski's humanist blockbuster emerged, this chapter compares *The Ascent of Man* (1973) with other BBC science features of the period, in particular Nigel Calder's *The Life Game* (1973). While the BBC's Adrian Malone (1937–2015) produced both of these shows, it was Bronowski's epic mode that got the biggest audiences, and in turn influenced a new wave of popular, progressive narratives about science. In addition to the narrative devices that helped make Bronowski's approach so popular, the chapter explores how the format and production techniques honed on *Ascent* directly influenced future big-budget science productions, both at the BBC (*Life on Earth*, 1979) and elsewhere (Cosmos, 1980).

The Ascent of Man

When Bronowski met with Singer and Reid in June 1969 about the "history of science" series, he picked up where *Insight* had left off, pitching three central concepts for the project:

1. that the series should "express the place of science in society";
2. that each episode would be dedicated to one scientific idea, so that the audience could "see the evolution of modern science along different pathways";
3. and that, reflecting his recent research on human specificity, each group of 3–4 episodes would focus on a different human faculty.

Singer and Reid knew that the lack of continuity between each episode of *Insight* had been a major factor in the series' limited success. The new history of science series was being modelled on *Civilisation*, which had successfully achieved narrative continuity by using a chronological structure. Following Singer and Reid's advice Bronowski agreed that two further concepts should ground the series:

4. the episodes will be arranged in a chronological sequence that demonstrates the progress of human societies;

Rethinking History, Science, and Religion: An Exploration of Conflict and the Complexity Principle, ed. Bernard Lightman (University of Pittsburgh Press, 2019).

5. and scientific ideas should be presented as the progressive solutions to major societal problems.[17]

For several years Bronowski had been wanting to distil his life's work into a single "intellectual autobiography", which could "create a philosophy for the twentieth century which shall be all of one piece".[18] He had always presumed this would be a written work, however he realised the BBC's offer presented him with the perfect opportunity. He could place the didactic case studies developed for *Insight* into a wider philosophical framework, which used historical case studies to make a progressive argument about the role of science in society. Since the 1940s, Bronowski had been developing his own version of scientific humanism (Chap. 3). His radio series *The Common Sense of Science* (1948) had utilised a sanitised version of the history of science and a chronological structure to place science within a progressive narrative of human knowledge, which included art and other forms of culture. Bronowski developed this argument further in a 1961 essay, "Science is Human", in which, capitalising on "the two cultures" debates sparked by C. P. Snow's 1959 Rede Lecture, Bronowski outlined how Humanism and the Scientific Revolution were intertwined parts of Renaissance activity in Europe.[19] Highlighting that scientific knowledge was only ever an approximation, Bronowski painted science as humanity's greatest cultural achievement, using a triumphalist history of the Scientific Revolution to make the case for the positive role of science in society. This essay was a more in-depth and philosophically informed exposition of the rationale Bronowski gave for the *Insight* series. In both, Bronowski grounded his synthetic vision, which incorporated everything from the formation of a star to an individual sneezing, in the explanatory mechanisms first provided by biological evolution. Since joining the Salk Institute in 1962, Bronowski had continued to develop this

[17] Correspondence from Jacob Bronowski to Robert Reid (Head of Science and Features, BBC TV), 28 July 1969, T64/64/1, BBC-WA. All BBC copyright content reproduced courtesy of the British Broadcasting Corporation. All rights reserved.

[18] Correspondence from Jacob Bronowski to Robert Reid, 28 July 1969; and Jacob Bronowski, *The Ascent of Man* (Little Brown, 1973), 17. All BBC copyright content reproduced courtesy of the British Broadcasting Corporation. All rights reserved.

[19] Julian Huxley edited the collection the essay appeared in, which grew out of Huxley's short-lived "Idea-Systems Group", a regular meet-up of friends and colleagues, which aimed to connect disparate scientific knowledge and provide "a new vision" of man's "destiny". Julian Huxley, *The Humanist Frame* (Allen & Unwin, 1961), 5–6.

line of research, forging an evolutionary version of the history of science, which he hoped could bridge the sciences and the humanities.[20]

In August 1969, Bronowski agreed to take on the BBC's history of science project on the condition that his own autobiographical account would be central, proposing the title "The Ascent of Man". The production team began the gargantuan pre-production process, drafting budgets, scouting locations across the world and fleshing out the content from Bronowski's initial outline. While Bronowski had agreed in principle that the episodes should be chronological, his original outline lacked a clear chronological sequence and overarching narrative. By early autumn 1969 the producer Adrian Malone, who had been the creative driving force behind the series since its first utterance, had begun crafting Bronowski's original synoptic outline into a more comprehensive production plan.[21] As part of this process, Malone tightened the chronology, and made the case, somewhat to Bronowski's chagrin, that without losing the original thematic ambition, "there should be another humanity besides that of Dr Bronowski".[22] Working closely with co-producer Dick Gilling, Malone reworked Bronowski's outline to ensure that the theme of "man's unique gift of imagination ran throughout", there was a "subtle chronological emphasis", and that each episode featured a historical figure who had made the "imaginative step".[23]

In these brief, innocuous pre-production exchanges, we see how the contingent process of creating broadcast media shapes the way that

[20] Robert Westman, 'The "Two-Cultures" Question and the Historiography of Science in the Early Decades of the Salk Institute for Biological Studies', 68.

[21] Bronowski and Malone had already collaborated on two "dramatised essays" one on Leonardo da Vinci and a second on the poet William Blake, both on BBC2, airing in early 1968 and 1969 respectively. See: "William Blake 1757–1827: As A Man Is – So He Sees", presented by Jacob Bronowski, directed by Adrian Malone, February 6, 1969, BBC2, and "Leonardo da Vinci", presented by Jacob Bronowski, directed by Adrian Malone, January 6, 1968, BBC2.

[22] For example in Bronowski's original outline, episode two was to chart the progress from hunter-gatherers to agriculture, and then on to plant and animal breeding up to genetics. In developing the series and script, Malone working with Bronowski, narrowed the historical scope of the episode to cover just the development of agriculture and the establishment of civilisations, moving the discussion of genetics to the series' penultimate episode. See folders T64/64/1 and T64/65/1 (BBC-WA) for the iterations and development of the series structure and script.

[23] Adrian Malone, "The Ascent of Man", BBC Internal Memo to Robert Reid (Head of Science and Features), September 22, 1969, T64/64/1, BBC-WA.

scientific content is communicated to the public. In October 1969, Singer put the final piece of the jigsaw in place as he, Reid and Bronowski met with Time-Life productions in New York securing a co-production agreement worth £420,000.[24] This extra funding enabled location shoots in over 20 countries, helping to ensure that the series that emerged was Bronowski's take on the evolutionary epic writ-large; the story of humanity depicted perhaps as grandly as it ever had been. Despite Bronowski's strong editorial control of the project, it was Malone's understanding of how to communicate science via television that shaped his more esoteric vision. By insisting on a more consistent chronology and including the great men of science, Malone ensured the series had a narrative structure familiar to audiences, shaping Bronowski's more general progressive outline into a more recognisable evolutionary epic.

The first episode of Bronowski's 13-part series, "Lower than the Angels" aired on BBC2 on May 5, 1973. In the opening sequence, Bronowski explicitly framed the narrative that was to follow:

> Man is a singular creature. He has a set of gifts which make him unique among the animals, so that unlike them he is not a figure in the landscape, he is the shaper of the landscape. ... [N]ature, that is evolution, has not fitted man to any specific environment ... he has a rather crude survival kit, and yet this is the paradox of the human condition: one that fits him to all environments. His imagination, his reason, his emotional subtlety and toughness make it possible for him, not to accept the environment, but to change it, and that series of inventions by which man from age to age has re-made his environment is a different kind of evolution; not biological, but cultural evolution. I call that brilliant sequence of cultural peaks, *The Ascent of Man*.[25]

[24] Of the originally agreed total of £423,800, the BBC were to cover 2/3 with Time-Life providing the final 1/3 in return for distribution rights in the US, with any profits shared equally. Head of Finance Services (Television), "Proposal for Increase in Expenditure", BBC Memo to Managing Director, Television, April 7, 1970, T62/1/1, BBC-WA; and Noble Wilson, "Time-Life co-production", BBC Memorandum, 3 September 1969, T64/64/1, BBC-WA.

[25] *The Ascent of Man*. "Lower than the Angels." Directed by Adrian Malone. Written by Jacob Bronowski. BBC/Time Life, 5 May 1973, BBC2: 00:07–02:40. This monologue ran prior to the title sequence, a technique known in television production as a "cold open", more common in this period in the US, and rarely deployed on documentaries; one notable exception released the same year as *The Ascent of Man*, which used a cold open on every

The episode began the series with a sweeping depiction of *Homo sapiens'* evolution. The viewer at home was left in no doubt about the scientific, humanist and, at points, atheistic story Bronowski was about to tell. Speaking from the East African Rift valley, he stated:

> The ancient stories used to put the creation of man into a golden age and a beautiful legendary landscape. If I were telling the story of Genesis now I should be standing in the Garden of Eden, but this is manifestly not the Garden of Eden, and yet I am at the navel of the world, at the birth place of man here in the East African Rift valley near the equator, and if this ever was a Garden of Eden, why it withered millions of years ago.[26]

It was the overtly secular humanist narrative of the series, exemplified by this swipe at organised religion, and Bronowski's significant editorial control of the script that led to the BBC adding the subtitle "A Personal View".

After beginning with proto-apes in the first episode, the series thematically jumped across the scientific ideas Bronowski thought had most transformed humanity, coming up to the present with an episode on genetics and cloning (Ep 12). While Bronowski's idea to deal with historical scientific developments thematically was still evident, Malone's work tightening the chronology and structure (Table 6.2) ensured *Ascent's* format and narrative arc were familiar enough for mainstream audiences. In situating humans within a grand evolutionary narrative, what five years later E. O. Wilson would call the "evolutionary epic",[27] Bronowski's progressive story of humanity aligned with a tradition that historians of science have subsequently traced back to the early nineteenth century.[28]

Following closely the format of *Civilisation*, and building upon the experiences of prior BBC science features, *Ascent* was a significant moment for science broadcasting. It took the communication of science on the BBC from studio demonstrations out into the field, utilising on-location

episode was: *The World at War*. Produced by Jeremy Isaacs. Narrated by Laurence Olivier. ITV/Thames Television, 1973–1974.

[26] *The Ascent of Man*. "Lower than the Angels." 07:10–07:53

[27] Edward O. Wilson, *On Human Nature* (Harvard University Press, 1978): 201.

[28] See, James A. Secord, *Victorian Sensation: The Extraordinary Publication, Reception, and Secret Authorship of Vestiges of the Natural History of Creation* (University of Chicago Press, 2003): 41–90; Bernard Lightman, *Victorian Popularizers of Science: Designing Nature for New Audiences* (University of Chicago Press, 2009): 219–223; and Ian Hesketh, "The Recurrence of the Evolutionary Epic," *Journal of the Philosophy of History* 9:2 (2015): 196–219, https://doi.org/10.1163/18722636-12341300.

Table 6.2 Episodes of *The Ascent of Man* and their constituent scenes

Episode	*Scenes*
1. Lower than the Angels	Shaper of the Landscape/The Rift Valley/Australopithecus/Physical Gifts/Precursors of Man/The hunter/The Ice Age/Cave Art
2. The Harvest of the Seasons	A Quiet Explosion/The Bakhtiari/Crossing the Bazuft/Wheat/Jericho/Agricultural Invention/The Horse/Genghis Khan
3. The Grain in the Stone	An Architect of his Environment/Splitting Stone/Machu Picchu/Pizarro/The Arch/Stonemasonry/The Sculptor/The Watts Towers
4. The Hidden Structure	Fire and the Alchemists/Copper/Bronze/Steel/Gold/Paracelsus/Oxygen/John Dalton
5. Music of the Spheres	A Harmony in Nature/Pythagoras' Theorem/Euclid and Ptolemy/The Coming of Islam/The Alhambra/Crystals/Toeldo/Perspective
6. The Starry Messenger	Easter Island/Wheels Within Wheels/Galileo/The Starry Messenger/Heresy/Pope Urban VIII/The Inquisition/The Verdict
7. The Majestic Clockwork	Newton/The Spectrum/Liebnitz/Elliptical Orbits/Three Hours After Marriage/Space and Time/Einstein/The Heart of Knowledge
8. The Drive for Power	The Industrial Revolution/The Marriage of Figaro/Benjamin Franklin/Iron/Wedgewood and the Lunar Society/Steam/Eternal Delight/Trevithick and Joule
9. The Ladder of Creation	Charles Darwin/Wallace and Bates/The Rio Negro/Wallace Returns/Survival of the Fittest/The Origin of Species/Louis Pasteur/Building Blocks of Life
10. World Within World	Salt/Mendeleev/Niels Bohr/Spectrum Analysis/Atomic Structure/Discovery of the Neutron/Evolution of the Elements/Immortality and Mortality
11. Knowledge or Certainty	The Method of the Artist/The Invisible Waves/Karl Friedrich Gauss/Göttingen/Born and Heisenberg/The Principle of Uncertainty/Leo Szilard/The Tragedy of Scientists
12. Generation Upon Generation	Gregor Mendel/The Garden Pea/All or Nothing/Sexual Reproduction/The Double Helix/The Fertilised Egg/Clones/Sexual Selection
13. The Long Childhood	The Plasticity of the Mind/The Brain/Speech/The Ability to Plan/Guardians of Integrity/John von Neumann/The Computer and the Brain/A Scientific Civilisation

"The Ascent of Man—Viewing Notes", BBC DVD/BBC Worldwide, 2005

shoots more familiar to audiences of natural history programming. A foundational humanist blockbuster, *Ascent*, had a direct influence on the subsequent emergence of "blue-chip" natural history films at the BBC and beyond.[29]

[29] Hall, 'A Humanist Blockbuster? Jacob Bronowski and *The Ascent of Man*'.

Unlike the earlier science features output in Chap. 3 and the educational content in Chap. 4, *Ascent* was not primarily about evolutionary science.[30] However, as with *Insight*, Bronowski situated the whole series within an evolutionary framework. The show's title was a play on Charles Darwin's 1871 *The Descent of Man*, and a knowing nod to the evangelist and biologist Henry Drummond who had first flipped the title in his 1894 publication *The Ascent of Man*. A Christian evolutionist, Drummond's account attempted to revise Darwinian evolution by replacing the emphasis on conflict with altruism; in Drummond's evolutionary epic, evolution was simply God's method.[31] Bronowski's series was a secular version of this progressive narrative, which made the bold assertion that the history of science was inextricably linked to the history of human evolution.[32] While Bronowski avoided the common, but incorrect, depiction of biological evolution as directional, the series' unashamedly progressive narrative of man's cultural evolution, a phenomenon he determined as completely separate from our biological evolution, on occasion slipped into the teleological.[33]

[30] For a later example focused on evolutionary science see *What is Life?* (BBC1, 1967), a ten part series directed by Mary Hoskins, and presented by British biochemist Professor Asher Korner, the series covered the latest understandings of molecular biology and featured a wide range of biological experts including the French biochemist Jacques Monod who appeared on three episodes.

[31] Bernard Lightman, 'Darwin and the Popularization of Evolution', *Notes and Records of the Royal Society*, 20 March 2010, https://doi.org/10.1098/rsnr.2009.0007, 17–20.

[32] Andrew Cunningham and Perry Williams, "De-Centring the "big Picture": The Origins of Modern Science and the Modern Origins of Science," *The British Journal for the History of Science* 26:4 (1993): 407–432, https://doi.org/10.1017/S0007087400031447.

[33] For more on teleology in television documentary see: Meryl Aldridge and Robert Dingwall, "Teleology on Television?: Implicit Models of Evolution in Broadcast Wildlife and Nature Programmes," *European Journal of Communication* 18: 4 (2003): 435–453, https://doi.org/10.1177/0267323103184001. For more recent discussion around teleology, in particular the argument that evolution can be directional without being teleological, see: George R. McGhee Jr., 'Can Evolution Be Directional without Being Teleological?', *Studies in History and Philosophy of Science Part C: Studies in History and Philosophy of Biological and Biomedical Sciences*, Special Issue: Replaying the Tape of Life: Evolution and Historical Explanation, 58 (August 2016): 93–99, https://doi.org/10.1016/j.shpsc.2015.12.006.

Only two of the thirteen episodes were on biology, both of which were almost exclusively about the history of evolutionary science.[34] Bronowski started "The Ladder of Creation" by stating that before the theory of evolution, the study of the physical world and the processes of life were separate paradoxical endeavours. Unlike many other popular iterations of the history of evolution, Bronowski included both Darwin and Wallace's roles in the co-discovery (Chap. 8), praising the apparently distinct cultural conditions of Victorian Britain that enabled two men to discover the same thing independently.[35] From Charles Darwin's family home, Down House, Bronowski introduced Darwin's foundational voyage on the *Beagle*, before giving a detailed biography of Wallace. *Ascent* brought Wallace's travels to life with dramatised excerpts from the journal of his journey up the Rio Negro; remarkably, this was the first time Wallace's story had been given substantial airtime on British television.[36] Informed by the latest research in the history of science, Bronowski returned to Down House and placed the inspiration for Darwin's imaginative leap to explain why species diverge on his reading of Thomas Malthus' *An Essay on the Principle of Population* (1798).[37] For Bronowski, the theory of evolution by natural selection was the most important scientific idea of the nineteenth century:

> Every generalisation about biology is a slice in time, and its evolution, which is the real creator of originality and novelty in the universe. And if that is so,

[34] Episode nine "The Ladder of Creation" and episode twelve "Generation Upon Generation".

[35] *The Ascent of Man*. "Ladder of Creation." Directed by Dick Gilling. Written by Jacob Bronowski. BBC/Time Life, 30 June 1973, BBC2: 01:30–04:10.

[36] Inspired by contemporary naturalists including Charles Darwin, Wallace left for Brazil in 1848 and would not return until 1852. For more on his journey see: Alfred Russel Wallace, *Travels on the Amazon and Rio Negro* (Ward Lock & Co., 1889), available online at https://archive.org/details/travelsonamazonr00wall/page/n9/mode/2up.

[37] In 1967, lost pages from one of Darwin's notebooks were published, leading to a revisiting of the contested influence of Malthus on Darwin's own work. For more see: Sandra Herbert, 'Darwin, Malthus, and Selection', *Journal of the History of Biology* 4, no. 1 (1 March 1971): 209–217, https://doi.org/10.1007/BF00356983; and Peter Vorzimmer, 'Darwin, Malthus, and the Theory of Natural Selection', *Journal of the History of Ideas* 30, no. 4 (1969): 527–542, https://doi.org/10.2307/2708609.

then each one of us traces back through that evolutionary process right to the beginnings of life.[38]

Bronowski removed any ambiguity with regard to the repercussions of Darwin's and Wallace's ideas, adding: "from that moment it was no longer possible to believe any story which supposed that at any time now there could be created once again the beginning of life".[39] Unlike educational broadcasts, which focused on teaching the mechanisms and processes of evolution, Bronowski emphasised the big picture and the societal repercussions of the theory. He stopped short of stating that evolution displaced God completely; such an overtly atheistic stance would have still been controversial for some of the desired primetime Saturday night audience. However, he was categorical in stating that evolutionary theory gave humans a clear impetus for progressive social change.[40] Like George Orwell's 1946 radio play (Chap. 5), by focusing on Darwin's own anxieties and relying on popular accounts of *Origin*'s impact, Bronowski's narrative centred on the revolutionary and conflict-centred societal repercussions of the theory.

The episode then moved to France and proceeded from the work of Louis Pasteur to the building blocks of life, amino acids. Following a scientific account of the likely origins of life, and the experiments of Bronowski's contemporaries, American chemist Stanley Miller and British chemist Leslie Orgel,[41] the episode closed by zooming out:

[38] *The Ascent of Man*. "Ladder of Creation." 30:05–30:41.

[39] *The Ascent of Man*. "Ladder of Creation." 31:52–32:05.

[40] In his less constrained written work, Bronowski did go further with this argument, arguing that there was no place for any kind of "magic" in human society. Jacob Bronowski, *Magic, Science, and Civilization* (Columbia University Press, 1978), 86–88.

[41] Leslie Orgel was a colleague of Bronowski's at the Salk Institute in California, where since 1964 he had been the director of the Chemical Evolution Laboratory, while Stanley Miller worked a short walk away in the Chemistry Department at the University of California at San Diego, where he was based from 1960. Following *Ascents* airing, Orgel and Bronowski featured together on BBC Radio 3 in August 1974 discussing their work on The Council for Biology in Human Affairs at the Salk Institute. Tom Blundell, 'Orgel, Leslie Eleazer (1927–2007), Theoretical Chemist and Molecular Biologist', Oxford Dictionary of National Biography, accessed 3 November 2020, https://doi.org/10.1093/ref:odnb/99309; Jeffrey L. Bada and Antonio Lazcano, *Stanley L. Miller (1930–2007): A Biographical Memoir*. National Academy of Sciences (USA), 9–10; and *Scientifically Speaking*, Presented by John Maddox, BBC Radio 3, August 12, 1974.

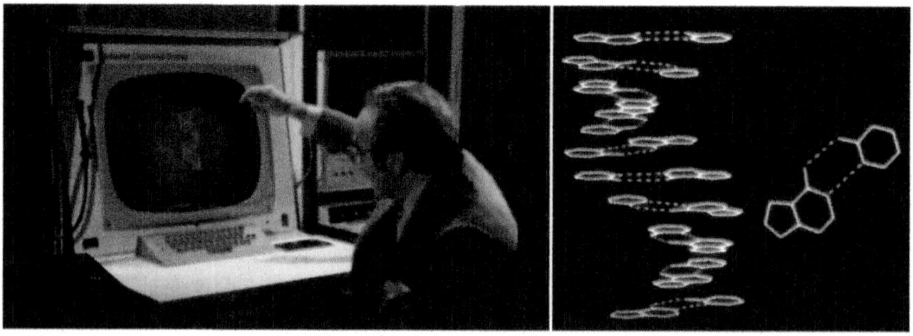

Fig. 6.1 (Left) Bronowski uses a computer to visualise Linus Pauling's incorrect three-helix structure of DNA as proposed by the structural chemist in early 1953, and (right) he uses the same computer to construct the correct structure as put forward by Watson and Crick in the spring of 1953. *The Ascent of Man*. "Generation upon Generation." Directed by Adrian Malone. Written by Jacob Bronowski. BBC/Time Life, July 21, 1973, BBC2: 28:24 & 31:37

> It may turn out that life had more varied beginnings, and has more varied forms. It doesn't at all follow that the evolutionary path which life, if we discover it elsewhere, took resembles ours. It doesn't even follow that we shall recognise it as life, or that it will recognise us.[42]

In finishing with this statement, Bronowski brought viewers back to his own personal thesis, situating neo-Darwinian evolution and the history of its discovery within his own cosmic evolutionary framework.

Returning to the biological in episode 12, Bronowski took the audience back to nineteenth-century Vienna, where against a backdrop of revolution, "the founder of genetics, and therefore the founder of all modern life sciences" Gregor Mendel was a student. After using time-lapse photography to illustrate Mendel's pea hybridisation experiments undertaken at his monastery in Brno, Bronowski recounted how Mendel's work was forgotten—erroneously inferring that his papers were burned because his work on hereditary was considered heretical.[43] Following an account of the science of sexual selection, Bronowski jumped forward into the twentieth century to cover the emergence of genetics. The first, and

[42] *The Ascent of Man*. "Ladder of Creation." 47:35–48:01.

[43] Edward Edelson, *Gregor Mendel: And the Roots of Genetics* (Oxford University Press, 2001), 13.

perhaps only, time that the structure of DNA has been outlined on prime-time British television, Bronowski used computer graphics (Fig. 6.1) to illustrate his account of Watson and Crick's discovery. As the scene progressed, the futuristic computer graphics were combined with short repetitive sounds, creating a kind of technological futuristic mantra, building to a crescendo as Bronowski's narration reached the breakthrough discovery of the structure of DNA.[44] The producers repeated this audio-visual tactic in the following scene, but now Bronowski was describing "DNA in action", and so, following microscopic imagery of egg fertilisation and embryo development, the scene this time crescendoed with unedited footage of a baby being born.[45] As the baby finally emerged, the accompanying soundtrack snapped to silence and the cries of the baby and the jovial conversation of the hospital delivery room could be heard. In breaking the rhythmic and at times tense audio-visual accompaniment with the sight and sounds of a newborn human, the scene attempted to disarm the unknown and perhaps unsettling nature of new genetic discoveries by connecting them to the most fundamental aspect of our shared humanity.

The episode next turned to cloning, with Bronowski contrasting bees' reproductive norms and social structures with humans. Making the case for the uniqueness of *Homo sapiens* using some spurious facts about our sexual behaviours—claiming humans were the only species to copulate face to face, and that such practice was universal in all cultures—Bronowski explored the "borderline" between the biological process of natural selection and the cultural pressures that affect sexual selection. He concluded, somewhat grandiosely, "we are uncommonly careful in the choice, not of whom we take to bed, but by whom we are to beget children. Sex was invented as a biological instrument by the blue-green algae, but as an instrument in the ascent of man which is basic to cultural evolution it was invented by man himself."[46] Here, as elsewhere in the series, when Bronowski switched to discussing more recent scientific discoveries, the

[44] *The Ascent of Man*. "Generation upon Generation." Directed by Adrian Malone. Written by Jacob Bronowski. BBC/Time Life, 21 July 1973, BBC2: 25:04–32:05.

[45] While one couple wrote in to the BBC stating that this episode alone justified their BBC licence fee, several others complained about the inclusion of the childbirth sequence. "The Ascent of Man – 12", BBC Internal Review, July 25, 1973 (211), 10, microfilm at BBC-WA. All BBC copyright content reproduced courtesy of the British Broadcasting Corporation. All rights reserved.

[46] *The Ascent of Man*. "Generation upon Generation." 48:46–49:15. All BBC copyright content reproduced courtesy of the British Broadcasting Corporation. All rights reserved.

mode changed from biographical and historical to explanatory and pedagogical. Thus, unlike the time afforded to discuss the societal repercussions of Darwin and Wallace's publications, no time was dedicated to the discussion of controversial issues in contemporary biology. While he conveyed scientific facts and knowledge in both the historic and the recent modes, Bronowski reserved the more expansive moments in the recent mode for his own thoughts, ideas and conjecture.

In the previous episode, "Knowledge or Certainty" Bronowski had discussed more explicitly the moral and ethical implications of scientific practice and discoveries. In particular, in the final two scenes in which Bronowski addressed scientists' role in the creation of the atomic bomb. The episode famously culminated with Bronowski standing in a tailings pond containing human ash at the former Nazi concentration camp of Auschwitz. Earlier in this again visually and audibly visceral scene, Bronowski recalled:

> I had not long been back from Hiroshima, when I heard someone say in [Leo] Szilard's presence, that it was the tragedy of scientists that their discoveries were used for destruction. Szilard replied, as he more than anyone else had the right to reply, that it was not the tragedy of scientists, it is the tragedy of mankind.[47]

This quote is indicative of other instances during the series, when Bronowski discussed moral dimensions. Despite many references to the messiness and contingency of science, and the connections between science and other cultural endeavours, ultimately Bronowski implored the audience to believe in a Baconian vision of science as value-free and neutral.[48] In the final episode, Bronowski stated, "knowledge is our destiny",

[47] *The Ascent of Man*. "Knowledge or Certainty." Directed by Mick Jackson. Written by Jacob Bronowski. BBC/Time Life, 14 July 1973: 43:25–44:00. All BBC copyright content reproduced courtesy of the British Broadcasting Corporation. All rights reserved.

[48] In his academic work Bronowski was engaged with the philosophy of science, most notably the work of Michael Polanyi, and Karl Popper who was a frequent visitor at the Salk Institute, thus he was familiar with the most contentious and unanswered philosophical aspects of contemporary evolutionary theory. See Jacob Bronowski, "New Concepts in the Evolution of Complexity: Stratified Stability and Unbounded Plans," *Zygon* 5:1 (1970): 18–35, https://doi.org/10.1111/j.1467-9744.1970.tb00181.x; and Robert Westman, 'The "Two-Cultures" Question and the Historiography of Science in the Early Decades of the Salk Institute for Biological Studies', 65.

and warned those in the West to "not retreat from knowledge", but to embrace our position as "nature's unique experiment to make the rational intelligence prove itself sounder than the reflex".[49] Much like Huxley's scientific humanism before him, for Bronowski the moral and ethical implications of science were for society as a whole to deal with. Bronowski's selective representation of Leo Szilard and more generally his other popular works promoting the redemptive power of science have been criticised by academics, who have noted the mythmaking and hypocrisy presented; especially given Bronowski's own personal involvement in ballistics and bombing during the war.[50]

Producing Ascent

A series the size of *Ascent* required a large production team, and as we have already glimpsed, a large number of experts shaped the final script, structure and production. Bronowski was assisted by his editorial assistant at the Salk Institute, Sylvia Fitzgerald, and his youngest daughter Judith Bronowski herself an educational filmmaker who in 1968 had made a film about the universe with a cosmic worldview.[51] On the BBC side, science features staff member Josephine Gladstone was responsible for co-ordinating research and liaising with external experts.[52] Once Malone and

[49] Jacob Bronowski, *The Ascent of Man* (Little, Brown, 1973), 437. All BBC copyright content reproduced courtesy of the British Broadcasting Corporation. All rights reserved.

[50] See Jerome R. Ravetz, *Scientific Knowledge and Its Social Problems* (Transaction Publishers, 1973): 65; David Edgerton, "British scientific intellectuals and the relations of science, technology, and war," in *National Military Establishments and the Advancement of Science and Technology: Studies in 20th Century History*, ed. Paul Forman and José Manuel Sánchez Ron (Springer Science & Business Media, 1996), 22–23; and Ralph Desmarais, 'Jacob Bronowski: A Humanist Intellectual for an Atomic Age, 1946–1956', *The British Journal for the History of Science* 45, no. 4 (December 2012): 573–589, https://doi.org/10.1017/S0007087412001069.s, "Jacob Bronowski." For a more personal and reflective account of the relationship between Bronowski's wartime work and his post-war promotion of scientific humanism, see *My Father, the Bomb and Me*, Written by Lisa Jardine, BBC Four, 26 January 2011.

[51] Victor K. McElheny, "Jacob Bronowski Is Dead at 66; Leading Popularizer of Science", *The New York Times*, August 23, 1974, 32; and Nasser Zakariya, *A Final Story: Science, Myth, and Beginnings* (University of Chicago Press, 2017), 269, 279–280.

[52] In earlier records from prior to her divorce, Gladstone is referred to as Jo Marquand. Bronowski thanks both Gladstone and Sylvia Fitzgerald in the introduction to the book of the series. In addition to various production roles for BBC TV science features, on occasion Gladstone also appeared in front of the camera, most notably presenting two episodes of

Gilling had set out a matrix for the series, defining the main themes, locations and visuals, the production team began the extensive research required to pull together such a wide-ranging documentary series.[53] Those consulted included university academics, institutions such as the British Museum and science journalists.[54] The author and historian of science Colin Ronan (1920–1995) carried out the research that informed much of the history of science content of the series, particularly the scientific biographies. The extent of this research reflected Bronowski's own desire to make the series as global in scope as possible. For example, he personally reached out to the USSR Academy of Sciences to ensure that Russian science received "adequate space and importance".[55] These attempts resulted in a series that was more global in its outlook and less hagiographical or "Whiggish" than many popular histories of science at the time. In episode one, Bronowski credited the Mayans for their sophisticated astronomy and mathematics, developed independently of the Old World.[56] However, throughout there was a tension between Bronowski's assertion that science was a "social enterprise" and the series' reliance on the usual cast of white, male, predominantly European scientists.[57] Although the prominence of famous historical individuals in the series came largely because of the demands of the format, even in Bronowski's original outline *Ascent* was a paternalistic account. In reducing the history of all knowledge to only one mode of knowing, science, the series presented an inherently progressive narrative of Western thought. As the historian of science Nasser Zakariya

Horizon in 1969 and 1970. "Master of the Microscope", *Horizon*, presented by Jo Marquand, October 20, 1969, BBC2; "King Solomon's Garden", *Horizon*, presented by Jo Marquand, February 9, 1970, BBC2; Jacob Bronowski, *The Ascent of Man* (Little, Brown, 1973); 'Ascent of Man (The), General 2', T64/65/1, BBC-WA; and "Mrs Jo (Marquand) Gladstone, The Ascent of Man", Jesus College Cambridge Archives, Bronowski Papers, 6/1/1.

[53] Adrian Malone, "The Ascent of Man", BBC Internal Memo to Aubrey Singer (Head of Features Group), September 30, 1969, T64/64/1, BBC-WA.

[54] See folders T64/64/1 and T64/65/1 at the BBC-WA.

[55] Jacob Bronowski, Letter to Millionshchikov (VP, USSR Academy of Sciences), June 9, 1970, T64/64/1, BBC-WA.

[56] *The Ascent of Man*. "Lower than the Angels." Directed by Adrian Malone. Written by Jacob Bronowski. BBC/Time Life, 5 May 1973: 04:29–04:50.

[57] *The Ascent of Man*. "The Drive for Power." Directed by Adrian Malone. Written by Jacob Bronowski. BBC/Time Life, June 23, 1973, 02:00–02:28, as quoted and discussed in Zakariya, *A Final Story*, 281.

has summarised, in *Ascent*, Bronowski effectively argued that science's "audience is universal, its great authors few".[58]

Ascent's filming schedule presented many logistical and geopolitical challenges, from ensuring one country would allow the crew entry after it had visited a rival territory, to making sure permits were in place for filming in protected areas. On occasion the actual content to be filmed was what blocked access.[59] After a request to the Congolese embassy for permission to film the Mbuti, pygmy hunter-gatherers indigenous to the rainforest of north-eastern Congo, Aubrey Singer received a letter denying the BBC permission to film in the country:

> If this report brings to science a contribution which is far from negligible it creates with the public a truncated image of the people. We know that listeners much enjoy the sight of what is original, primitive and wild, but we also know that the public which is not always well advised confuses the particular with the general. We cannot thus contribute to anything which may encourage the belief that the primitive Mbuti people still reflect the physionomy [SIC] of the Congo people who have suffered traumas and who have lost their dignity in the international scene. ... You will easily understand that this retreat towards the primitive is inopportune and that while it may allow people to see the splendour of the country [it] can alter other human values.[60]

This disagreement again highlights the tension of the series' attempts to tell a universal history of humanity through the lens of science, a predominantly Western endeavour historically implicated in racist colonial expansion. Indeed, while it was in keeping with most popular narratives of empire in the early 1970s, the omission of colonial histories of occupation in *Ascent* was at odds with the direction of travel in disciplines such as

[58] Zakariya, 281.

[59] By the early 1970s although occasional issues still arose, the process of securing filming permissions had been incorporated into the standard pre-production process of BBC TV Features. See folders T64/64/1 and T64/40/1 (BBC-WA) for discussion about whether the *Ascent* crew would be granted permission to film in China following a planned shoot in Taiwan. For an earlier instance of the challenges of gaining access to film in unstable and geopolitically sensitive areas see David Attenborough's account of trying to get to the Aru Islands in Indonesia to film birds of paradise for *Zoo Quest* in 1957: David Attenborough, *Life on Air: Memoirs of a Broadcaster* (Princeton University Press, 2002), 64–70.

[60] Joseph Kalume, Letter to Aubrey Singer, undated, T64/64/1 BBC-WA.

anthropology, literary studies and philosophy, where postcolonial and subaltern approaches were already emerging.[61]

A Popular Success, the Reception of Ascent

Aired in a primetime Saturday evening slot on BBC2, *Ascent* received plaudits from the popular press and the public at large, with one BBC report estimating that 30% of the adult viewing population of the UK, some 10–12 million people[62] saw at least one episode of the series during 1973–1974.[63] The BBC widely syndicated the series and it aired on television stations across the world, from Sweden to Australia. The BBC accompanied the show's launch with a co-ordinated media campaign.[64] Preceding a repeat run in November 1973, Bronowski appeared on the chat show *Parkinson*, and the book of the series was launched in December 1973.[65]

In general, the national press received the series well,[66] and in due course Bronowski was rewarded, winning the 1973 Writers' Guild Award for best documentary script, and a silver medal at the 1974 Royal Television Society Awards.[67] Praise was lauded not only on Bronowski, but also on the "superb professional talent" in the Features Department, including

[61] For early examples from the period, which predate the emergence of the academic field of postcolonial studies see: Frantz Fanon, *The Wretched of the Earth* (Grove Press, 1963); and Edward Said, *Joseph Conrad and the Fiction of Autobiography* (Harvard University Press, 1966). For an introduction to some of the ways that the history of science has, and can, incorporate postcolonial, transnational and global historical perspectives see: Kapil Raj, 'Beyond Postcolonialism … and Postpositivism: Circulation and the Global History of Science', *Isis* 104:2 (2013): 337–347, https://doi.org/10.1086/670951.

[62] In 1971 the total population of the UK was 55.9 million, with 25.5% estimated to be under 16, thus 30% of the over 16 population is approximately 12.49 million. Jen Beaumont, "Population," *Social Trends 41*, Office for National Statistics Report (Crown Copyright, 2010).

[63] The show originally aired on Saturday evenings from May to July 1973, and was repeated on Thursday evenings from November 1973 to January 1974. "What the Ascent of Man Conveyed to Viewers," An Audience Research Report, May 1974, T62/1/1, BBC-WA.

[64] See, for example, Jacob Bronowski, "Lower than the Angels – the first of 13 essays on man's relation to nature," *The Listener*, 10 May 1973; and *Whatever You Think*. Presented by Cliff Michelmore. 29 April 1973, BBC Radio 4.

[65] *Parkinson*. Directed by Brian Whitehouse. Presented by Michael Parkinson. 3 November 1973, BBC London. For a full transcript of this interview, see T64/64/1, BBC-WA.

[66] See for example Matthew Coady, "Mirror TV: Your Complete Programme Guide," *The Mirror*, 21 May 1973, 18.

[67] "Television Today: Writers' Awards," *The Stage*, 16 May 1974, 13; "Television Today: Peter Willes wins RTS award," *The Stage*, 30 May 1974, 13.

the producers Adrian Malone and Dick Gilling, the innovative graphics and filming techniques, and the BBC itself, the only "broadcasting organisation in the world which could have undertaken it and succeeded".[68]

Unsurprisingly, given the occasional digs at religion—"the Bible is a curious history, part folklore, part record",[69]—the BBC did receive some complaints about Bronowski's depiction of religion and belief. For example, a person who overall was enthusiastic about the series complained of Bronowski's "anti-religious dogmatism".[70] Echoing the format of the final episode of 1958s *Five Hundred Million Years*, in response in February 1974 BBC Radio 3, invited Bronowski on air to be questioned about his treatment of religious experience in the series by the Christian evolutionist Derek Burke (1930–2019).[71] Despite Burke's concerns, the British religious press made little comment on the series, and where they did review *Ascent*, criticism centred on its narrative and visual style.[72] More notable than the paucity of comment in the religious press was a complete lack of mention of the series in the most popular humanist magazine of the day, *The Freethinker*.[73] Despite this, the humanist overtones of the series did not go completely unnoticed, with *The Times'* referring to the final episode as a "humanist epilogue".[74]

Audiences, like the media, positively reviewed both the content and the ambitious presentation. However, BBC audience research reports reveal that most viewers interpreted the series simply as a history of man and science, rather than Bronowski's own objective to create "a philosophy for

[68] Chris Dunkley, "The Ascent of Man," *The Financial Times*, 7 May 1973. See also, Peter Lennon, "Man alive," *Sunday Times*, 6 May 1973, 38.

[69] *The Ascent of Man*. "The Harvest of the Seasons." Directed by Adrian Malone. Written by Jacob Bronowski. BBC/Time Life, 12 May 1973: 26:49–27:03.

[70] "What the Ascent of Man Conveyed to Viewers," An Audience Research Report, May 1974, T62/1/1, BBC-WA, 8.

[71] "Bronowski on God", Chaired by Robert Reid, BBC Radio 3, February 26, 1974. Derek Burke was Professor of Biological Sciences at the University of Warwick who often wrote about the relationship between his faith and evolutionary science. For an introduction to his views on the subject, see his chapter "Creation and Evolution" in: Derek Burke (eds), *When Christians Disagree: Evolution* (IVP-UK, 1985).

[72] Mary Crozier, "Television," *The Tablet*, 2 June 1973, 13.

[73] Here, as with other sectors of the press, we see highlighted the challenges for historians studying the reception of television in the period; often not considered highbrow enough, or considered too ephemeral to bother covering (why review a programme that, unlike a book or theatre performance, the reader can no longer access).

[74] "Plenty of time to grow up," *Sunday Times*, 22 July 1973, 52–53.

Fig. 6.2 Punch magazine cartoon criticising the overly simplistic narrative arc of *The Ascent of Man*. Bernard Hollowood, "Ascent my foot!" *Punch*, May 16, 1973. Reproduced with permission of Punch Ltd., www.punch.co.uk

the twentieth century, bringing together the experience of the arts and the discoveries of science". Aside from individual gripes centred on Bronowski's perceived anti-religious stance, the greatest concern of those surveyed was that the series was too intellectual.[75] Likewise, while letters in the press criticised aspects such as Bronowski's chauvinistic overtones, through to the historical accuracy of particular scenes, few, if any raised concerns about the religious implications of his worldview.[76] In reviewing the series the BBC were interested in knowing how viewers' prior beliefs affected their perception of the series. Despite an unsubstantiated expectation that "religious allegiance" declined with increased education, and some fairly flimsy results, the Audience Research Department concluded that belief *had* played a significant role in whether individuals tuned in to watch the series or not.[77]

[75] "What the Ascent of Man Conveyed to Viewers," An Audience Research Report, May 1974, T62/1/1, BBC-WA; "The Ascent of Man: A Personal View by J. Bronowski: 1," BBC Audience Research Report, 25 May 1973, BBC-WA; and "The Ascent of Man: A Personal View by J. Bronowski: 7," BBC Audience Research Report, 10 July 1973, BBC-WA.

[76] See respectively John Rowan, "Letters – 'The Ascent of Man'," *The Listener*, 16 August 1973; John Russell, "Letters – 'The Ascent of Man'," *The Listener*, 21 June 1973; and Bronowski's response to these criticisms: Jacob Bronowski, "Letters – 'The Ascent of Man'," *The Listener*, 23 August 1973.

[77] "What the Ascent of Man Conveyed to Viewers," 1–5.

The main criticisms levelled at the series focused on the oversimplified historical narrative and the overreach of what science can tell us about society and the physical world. This argument was made in both popular outlets, such as *Punch* magazine (Fig. 6.2), and academic forums, like the *American Anthropologist*, which complained of the "ubiquitously naïve ethnocentrism" and disparagingly referred to the series as an advertisement for culture.[78] These critiques reflect the tensions already highlighted; Bronowski's desire for a universal account and the requirements of the emergent primetime format left little scope for nuance or complication in the series' narrative arc. In many regards, Bronowski was a dinosaur, a throwback to a generation of science communicators who wished to speak in universal narratives using an authoritative tone.[79] As early as 1959, members of the BBC General Advisory Committee had questioned his suitability for television, and during an internal review of *Ascent* the head of television school broadcasting referred to Bronowski as "unctuous and condescending".[80] Indeed, the series was out of step with most BBC education and science features output in the early 1970s. However ultimately, this did not matter. When hitched to Malone and Gilling's exciting new production techniques, Bronowski's approach resonated with a large proportion of the British public, introducing a new generation to the evolutionary epic. The series and the subsequent book, which was still on best-seller charts in late 1975,[81] retained a continued popularity and went on to influence future non-fiction television programming, the formative field of science communication, and in turn, popular narratives on scientific progressivism. Before we return to the widespread and lasting legacy of *Ascent*, I will now introduce other approaches to communicating evolution broadcast in the early 1970s.

[78] Roy Wagner, "Television Productions: The Ascent of Man," *American Anthropologist* 79 (1977): 993–994.

[79] The historian of science Robert Westman argues that Bronowski's *Ascent* can be seen as a late expression of the trend for histories of science to be written by scientists, rather than professional historians of science. Although this view, somewhat overlooks the input during the production of *Ascent* of experts, including the historian of science Colin Ronan. Robert Westman, 'The "Two-Cultures" Question and the Historiography of Science in the Early Decades of the Salk Institute for Biological Studies', 83.

[80] Stafford Beer, *Minutes of the General Advisory Committee*, 1959, 26–28, as copied in T16/623, BBC-WA; and "The Ascent of Man: 1: Lower than the Angels", BBC Internal Review, May 9, 1973 (133), 8, microfilm at BBC-WA.

[81] "Best-Sellers," *Sunday Times*, 28 September 1975.

Other Popular Depictions of Evolution in the 1970s

In the late 1960s and early 1970s, "anti-science" sentiment was on the rise and the scientific establishment became increasingly worried about critical attitudes towards science (Chap. 5).[82] Alongside these concerns, during the middle decades of the twentieth century big syntheses and universal narratives, such as those favoured by Julian Huxley, fell out of favour.[83] Many television and radio features, both fiction and non-fiction, reflected these concerns. In addition to science-fiction such as *Doomwatch*, the early 1970s saw these concerns covered in regular science programming (e.g. *Horizon*), as well as in new experimental approaches, which attempted to allow more critical reflection on science's role in society.[84] One such show, *Controversy* (1971–1975) picked up the pre-war science and society thread, and put scientific experts in the firing line of a live-studio audience. Filmed in the historic lecture theatre at the Royal Institution in London, following a short talk the expert was debated by a panel of scientists selected for their opposition to the main speaker's ideas, before discussion was opened up to the audience. The format promised to give a popular forum for controversial scientific topics, which were conspicuous in their absence in *Ascent*. However, rather than the debate covering a cross-section of societal views, often discussion was dominated by specific interest groups in attendance. The show inherently prioritised scientific expertise over other forms of knowledge, and the dynamic of a rebellious audience attacking the scientific elite was further exacerbated by limiting the panel predominantly to scientific experts.[85]

[82] Rupert Cole, '1972: The BBC's Controversy and the Politics of Audience Participation', *Public Understanding of Science* 26, no. 4 (1 May 2017), https://doi.org/10.1177/0963662516684231, 515.

[83] Zakariya, *A Final Story*, 275.

[84] From its first season onwards *Horizon* (1964–present) regularly featured episodes in a critical mode, covering topical issues in relation to science, however it rarely, if ever, provided a space for criticism of scientific process itself, or the superior authority of scientific knowledge. See for example the second episode of the first series, "Pesticides and Posterity" (May 30, 1964, BBC2) as discussed and analysed in Timothy Boon, '"The Televising of Science Is a Process of Television": Establishing Horizon, 1962–1967', *The British Journal for the History of Science*, 48:1 (2015): 108–111. For an example from the early 1970s that illustrates how *Horizon* attempted to cover anti-science trends in society, see "Worlds in Collision" (January 11, 1973), which focussed on the popular pseudoscientific theories of Immanuel Velikovsky. The episode can be watched online https://archive.org/details/wrldsc *(last accessed 13 Nov 20)*.

[85] Cole, '1972'.

Of 29 episodes, six episodes had evolutionary themes.[86] One of these six, "A New Look at an Old Animal", was the most controversial of the series' inconsistent four-year run. In the episode, anthropologists Robin Fox (b.1934) and Lionel Tiger (b.1937) presented their argument that human behaviours, in particular the divergence of societal roles for men and women, were predominantly governed by evolutionary instincts acquired in prehistoric times.[87] The socialist feminist Juliet Mitchell (b.1940) had been included on the panel, in the expectation that she would be well placed to criticise their thesis.[88] However, the limelight was stolen by members of a women's liberation group in the audience, who vociferously critiqued Fox and Tiger's position, calling out their patriarchal language and the biases in their male-dominated account. Although most of the audience cheered the women's liberation members as they questioned Fox and Tiger, in a reminder that counter-cultural movements were just one view amongst many, the overwhelming majority of complaints following the episode targeted the women who disrupted proceedings.[89]

The week after this incendiary episode, *Controversy* again covered evolutionary themes as the French biochemist, Jacques Monod (1910–1976) outlined the main neo-Darwinian arguments of his best-selling book *Chance and Necessity* (1970).[90] Monod examined the philosophical implications of molecular biology, introducing a teleonomic argument that the

[86] The episodes were: "Race, Intelligence and Education", Professor Hans Eysenck, 23 Aug 1971; "A New Look at an Old Animal", Professor Robin Fox and Professor Lionel Tiger, 4 Sept 1972; "Chance and Necessity", Professor Jacques Monod, 11 Sept 1972; "Certain types of genetic research should be suspended", Professor Paul Berg, 16 Sept 1974; "A doctor has no duty to prolong life at any cost", Dr John Lorber, 27 Feb 1975; and "A discussion on child sterilisation", Special (no single proposer), 16 Aug 1975.

[87] Lionel Tiger and Robin Fox, *The Imperial Animal* (Dell Publishing, 1971).

[88] Although criticised by academics from both anthropology and the social sciences *The Imperial Animal* proved popular among non-academic audiences during the early 1970s. See: Alexander Alland, "Physical: *The Imperial Animal.* Lionel Tiger and Robin Fox." *American Anthropologist*, 75, 1147–1148. https://doi.org/10.1525/aa.1973.75.4.02a01600; and Margaret Sery Young, "*The Imperial Animal* by Lionel Tiger, Robin Fox", *Contemporary Sociology*, 1976, 5:3, 382–384. http://www.jstor.org/stable/2064160.

[89] For a more detailed account of the exchange that occurred during this episode see Cole, '1972'.

[90] In its first year, the book sold over 200,000 copies and was a best-seller in Germany and Japan. Oren Harman, 'Chance and Necessity Revisited', *Journal of the History of Biology* 47, no. 3 (1 August 2014): 479–493, https://doi.org/10.1007/s10739-014-9390-3, 479.

purposefulness of biological processes can be explained solely by the necessary, but choiceless interactions of different levels of biological systems. While today Monod's account seems overly reductionist, at the time, unlike Fox and Tiger, his thesis was not particularly controversial among his fellow biologists. Criticism of *Chance and Necessity* in the period came more from philosophers, who thought Monod was attempting to derive values from facts, and theists, who disagreed with his purely materialistic account.[91]

Although the Monod episode was rather tame in comparison to the week before, both episodes prompted senior executives to reflect on the limitations of the format.[92] *Controversy* aimed to give space for open discussion about science's role in society, but critique did not just focus on specific scientific theories, or those areas where there was a clear political dimension. The format, which gave a platform to a usually fringe scientific idea and pitched it against a panel of more mainstream positions, allowed the black box of scientific process to be opened in public. Further, interjections allowed an un-vetted audience to critique science on non-scientific terms, and for "transgression of the public–expert dichotomy".[93] This

[91] While biologists criticised specific parts of Monod's book, overall his popular account of contemporary molecular cybernetics and the implications for human society was welcomed. For examples of some of the most common responses to Monod's book see: Harman; John Maynard Smith, "Introduction to the Penguin Edition", in Jacques Monod, *Chance and Necessity* (Penguin Books, 1997), xi–xvi; F. Eugene Yates and Arthur S. Iberall, 'Chance and Necessity: An Essay on the Natural Philosophy of Modern Biology by Jacques Monod', *Annals of Biomedical Engineering* 1, no. 3 (1973): 381–384, https://doi.org/10.1007/BF02407677; Jeffrey S. Wicken, 'The Cosmic Breath: Reflections on the Thermodynamics ofCreation', *Zygon* 19, no. 4 (1984): 487–505, https://doi.org/10.1111/j.1467-9744.1984.tb00943.x; and Bertram Morris, 'Jacques Monod, Chance and Necessity: An Essay on the Natural Philosophy of Modern Biology (Book Review)', *Technology and Culture;* 13:4 (1972): 662–663.

[92] Issues raised during BBC internal reviews of these two episodes included the disparate nature that discussion was allowed to take, the lack of control the BBC had over infiltration of the audience by special interest groups and the decision to base discussions on a single book that most of the audience at home were unlikely to have read. "Controversy: A New Look at an Old Animal", BBC Internal Review, September 6, 1972 (240), 7; and "Controversy: 3: Chance and Necessity", BBC Internal Review, September 13, 1972 (246), microfilm at BBC-WA.

[93] Cole, 517. For more on the concept of black boxes in science and technology studies see: Bruno Latour, *Science in Action: How to Follow Scientists and Engineers through Society* (Harvard University Press, 1987), http://archive.org/details/scienceinaction0000unse, 1–13; and Massimiano Bucchi and Brian Trench, *Handbook of Public Communication of Science and Technology* (Routledge, 2008), 70–71.

critical approach that deconstructed the methods and processes of science, rather than just the application of its knowledge, was at odds with Bronowski's *Ascent*.[94] While *Ascent* was more tempered and reflective than the original optimistic tone of Huxley and Mary Adams' science and society approach, it was in essence a defence of science, placing responsibility for recent scientific and technological ills at the hands of those who misused scientific knowledge.

From this period of experimentation in science broadcasting, we can identify a third approach to television features on biological evolution. Not the triumphalist scientist's view of *Ascent*, nor the critical sociological turn of science studies that influenced *Controversy*, but a more expository dialectic style influenced by science journalism.[95] Spearheading this approach in the late 1960s and early 1970s was the science writer Nigel Calder (1931–2014). After a degree in the natural sciences, Calder followed in the footsteps of his science journalist and radio broadcasting father, Ritchie Calder (1906–1982), to become a founding writer for New Scientist in 1956, where he served as editor from 1962 to 1966, before going freelance.[96] Popular science writing and science journalism had long been a source of content for science on BBC radio.[97] Reflecting his background, Calder's approach to televised science was something of a conciliatory middle way; less revolutionary than the most progressive elements of the radical science movement, and less sycophantic than the most reductive narratives of progress and greatness.

[94] Timothy Boon, *Films of Fact: A History of Science in Documentary Films and Television* (Wallflower Press, 2008) 235–241.

[95] Following the appointment of J. G. Crowther (1899–1983) as the first scientific correspondent for the *Manchester Guardian* in 1928, science writing in the UK continued to grow in popularity and by the middle of the century had become a specialised and largely professionalised field, see: Jane Gregory and Steven Miller, *Science In Public* (Basic Books, 2000).

[96] Tim Radford, 'Calder, Nigel David McKail Ritchie (1931–2014), Journalist, Author, and Broadcaster', Oxford Dictionary of National Biography, https://www.oxforddnb.com/view/10.1093/odnb/9780198614128.001.0001/odnb-9780198614128-e-108856.

[97] One example of the direct influence of popular science writing on radio in the mid-twentieth century is how Ritchie Calder's *Profile of Science* (Allen & Unwin, 1951) served as the direct blueprint for a series proposed by Nesta Pain in 1951. Although in this instance the series never saw the light of day, because Ritchie Calder was out of the country for much of 1951, it is one of many examples in the archives of radio content that directly replicated, reviewed or was inspired by popular science writing in the period. See: Nesta Pain, "Proposed Light Programme series based on Profile of Science by Ritchie Calder", BBC Memo to Assistant Head of Features, June 11, 1951, R19/1094, BBC-WA.

Unlike *Ascent*, Calder's shows did not attempt to construct a universal narrative for the sciences. Yet at the same time, they also did not challenge the general progressive narrative of modern scientific activity. Rather, by focussing in-depth on the contemporary state of knowledge in specific fields, Calder zoomed in on minor episodes in this greater grand narrative. With his journalistic style and focus on the contemporary, crucially, Calder's broadcasts were not as introspective as Bronowski's approach in *Ascent*. While by design, *Ascent* looked backwards at the history of scientific achievement, in Calder's first major BBC commissions the focus was on cutting-edge ideas and future developments.[98] Although they were one-off specials, Calder's global approach, like *Ascent*, required a big budget and a long lead in time. Alongside his documentaries, Calder often worked freelance on other BBC productions. In the early planning of *Ascent*, he was brought in to assist Malone and Gilling with a preliminary scoping survey.[99] Building on this tentative collaboration, in 1971, once Malone had finished the bulk of filming and production for *Ascent*, he was tasked with producing Calder's next in-depth exploration, which was this time to tackle the "vast" content of contemporary biology.[100]

[98] *The Violent Universe*, presented by Magnus Magnusson and Ian Roxburgh, narrated by John Stockbridge, and written by Nigel Calder, BBC2, April 17, 1969; "The Mind of Man", presented by Magnus Magnusson, narrated by Marius Goring, and written by Nigel Calder, BBC2, October 23, 1970; and "The Restless Earth", presented by Magnus Magnusson, narrated by John Gregson, and written by Nigel Calder, BBC2, February 16, 1972. The Science Spectacular series were all co-productions with amended versions airing across several countries. The US version of *The Violent Universe*, presented by Carl Sagan aired on National Educational Television stations across the USA as part of the Public Broadcast Laboratory project, can be watched online at: https://archive.org/details/theviolentuniversepart1 *(last accessed 17/06/2021)*.

[99] Humphrey Fisher, "The Story of Science", BBC Internal Memo to Head of Science & Features, Television, January 31, 1968, T64/64/1, BBC-WA.

[100] Malone was only brought in after the promotion of the original producer and long-time Calder collaborator Philip Daly in early 1973. Malone had the unenviable task of joining the project after Calder and Daly had completed all of the research via an extensive reconnaissance trip visiting somewhere in the region of 100–200 academics in over ten countries. Nigel Calder, "Preliminary proposal for a two-hour television documentary", March 1972, T63/69/1, BBC-WA; and Nigel Calder, 'Science as Intellectual Entertainment', *Nature* 245, no. 5424 (October 1973): 293–295.

Nigel Calder's The Life Game

Just a few months after the final episode of *Ascent*, on the evening of Saturday October 13, 1973, *The Life* Game was broadcast on BBC2. Like Calder's earlier documentaries, *The Life Game* was a two and a half hour, one-off special, which promised "[a]n exciting new account of evolution in action" and introduced viewers to the latest neo-Darwinian developments in the biological sciences.[101] *The Life Game* was forward-looking; however, given the phylogenetic nature of our evolutionary development, it inevitably had to look backwards in some regards. In developing the content, Calder had considered biology too large a subject to tackle in one documentary, but in-line with Bronowski's approach in *Ascent* he considered evolution by natural selection a "great unifying theme", which made it manageable.[102]

Calder's usual presenter Magnus Magnusson was overlooked, as a now freelance David Attenborough was considered to be a more recognisable and authoritative figure when it came to the subject matter.[103] Linking the filmed sequences together from the studio, Attenborough began by reassuring viewers about any worries over their lack of knowledge; putting them in the same category as a hypothetically back-from-the-dead Darwin, who equally would not understand the jargon of the new biology.[104] With a nod to Jacques Monod's *Chance and Necessity*, Calder built the whole documentary around the extended metaphor of the history of life on earth as a game of chance[105]; albeit, an extremely long, high stakes and low

[101] Even those commentators who in general supported Calder's highbrow approach to documentary and elsewhere argued that the one-off in-depth treatment suited some subject matter, found *The Life Game* hard work. For example see: Chris Dunkley, "Real Television", *The Financial Times*, October 17, 1973, 3; and Chris Dunkley, "Blockbusters", *The Financial Times*, November 27, 1974, 3.

[102] Nigel Calder, "Preliminary proposal for a two hour television documentary", 1.

[103] Philip Daly, Letter to Magnus Magnusson, January 2, 1973, T63/69/1, BBC-WA.

[104] Nigel Calder, *The Life Game: Evolution and the New Biology* (The Viking Press, 1973), 9.

[105] Calder and Daly had originally set out to explore what was wrong with Monod's philosophy of molecular evolution, and while chance and probability in evolutionary processes featured heavily throughout the programme, they were no great fans of Monod's thesis, referring to *Chance and Necessity* as "still following the obsolete ideas of mutationism". Calder, *The Life Game; Evolution and the New Biology*, 64; and Calder, 295. While mutationism has changed and developed in meaning over the 120 plus years the term has existed, here Calder was using it to refer to the idea that mutations were the main driver of variation in species and that they were most often aberrant and disadvantageous. By the 1940s, this view was being rendered redundant by studies that focused on the genetics of populations rather than ideal types, and proponents of the evolutionary synthesis including Julian Huxley positioned mutationism outside of the neo-Darwinian synthesis. See John Beatty, 'The Creativity

probability game of chance.[106] While the traditional narrative touchstones remained, for example Darwin's voyage on the *Beagle* and the Galapagos Islands, the documentary also highlighted new research in evolutionary biology. For example, genetic explanations of altruism and the "great evolutionary heresey" of Motoo Kimura's (1924–1994) neutral theory of molecular evolution, published in 1968.[107]

of Natural Selection? Part II: The Synthesis and Since', *Journal of the History of Biology* 52:4 (2019): 705–731, https://doi.org/10.1007/s10739-019-09583-4.

[106] Although Table 6.3 outlines the chapters and sub-sections of the subsequent more in-depth book released alongside the documentary it gives a good indication of the type of language favoured by Calder in his framing of the latest evolutionary sciences.

Table 6.3 Overview of the chapters and sub-sections of the book version of *The Life Game*

	Chapter	*Sub-sections*
1.	Messages from the Dead	A ploy out of Africa/The molecular alternatives; History by hereditary/Our changeable skins; Mankind dispersed/What is the life game?
2.	How to Evolve	Identikit for an ancestor/The changeling gulls/Enchanted islands/The dire opponent/Rules and players/Nature abhors perfection/The theory of games/Thou shalt not kill/Nothing evolves alone.
3.	The Molecular Heresy	Unbridled genes/Doing without fossils/Clockwork mutations/Pockets of conservatism/Hereditary roulette/The evolution of evolution/Too many mutations/Symptoms of randomness/Darwinian backlash/An indelible mark.
4.	Grandmother Hypercycle	Recipe for soup/Dice and cloverleaves/An adaptable molecule/The contest of the hypercycles/Basic characters/A change of menu/Modernising the cells/Sex becomes interesting/Why mince hydra?/Instructions for Growth/The genetic computer.
5.	Visible Players	Fanfare for the worms/Coalescence and break-up/Planetary scene-shifting/A partial record/Country cousins/Rifts among the apes/Leakey's luck/Sociable hunters/Dead ends/Successors for man?
6.	Postscript: A Game of Skill Index	New responsibilities/New organisms/New people.

Nigel Calder, *The Life Game: Evolution and the New Biology* (The Viking Press, 1973)

[107] Calder, 'Science as Intellectual Entertainment', 293.

Although the life game metaphor featured in Calder's original outline, it was while travelling the world researching the documentary that it grew into the central theme. Reflecting how extensively mathematical work on game theory was being applied across the biological sciences during the early 1970s, the life as game metaphor was repeatedly raised in Calder's interviews with leading scientists.[108] While at the level of metaphor, games were something relatable to a general family audience, biological research that utilised game theory was complex. It was reliant on complicated probabilistic mathematics and laboratory techniques normally beyond the comprehension of the documentary's target audience. *The Life Game* attempted to overcome this barrier in three ways: Calder's accessible journalistic scripting, the use of visual graphics depicting abstract processes and extensive expository dialogue and demonstrations by David Attenborough in the studio. With regard to these last two elements, Malone and his team played a crucial role, coming up with creative solutions, such as a molecular fruit machine used to illustrate the difficulty of the first life emerging from a primordial soup.[109]

Unlike Bronowski, who came to *Ascent* with his thesis already formed, Calder, purposefully wanted his process of visiting hundreds of experts to be reflected in the final film. The result was a documentary that was "a process of discovery rather than a mass of schoolroom facts". Calder's approach and his focus on the "frontier" allowed the audience a small glimpse into the processes of science.[110] There was space for scientific disagreement, competing theories and changes in direction; viewers could peek inside the black box of biological research, but unlike in *Controversy*, any critique remained purely on scientific terms.

The Life Game was a seven-way co-production, carefully constructed so that it could be easily tailored to suit each broadcaster's national context.[111] To date, it is perhaps still the most successful attempt at

[108] Nigel Calder, "Preliminary proposal for a two-hour television documentary"; and Calder, 'Science as Intellectual Entertainment'.

[109] In his review of the documentary in *The Financial Times*, critic Chris Dunkley singled out the purpose-built studio fruit machine as an indicator that the documentary had been too well funded. Chris Dunkley, "Real Television", 3.

[110] Calder, 293.

[111] The seven broadcasters were the BBC, WNET/13 (USA), Sveriges Radios (Sweden), ABC (Australia), BRT/KRO (Belgium/The Netherlands), Studio Hamburg (Germany) and the Ontario Educational Communications Authority (Canada). Earlier partners JBC/NHK (Japan) pulled out during the production process. While all of the studios recorded narration

communicating the full complexity of genetics and molecular biology to a primetime audience. However, in this instance, Calder's skill for clear exposition and Malone's talent for audio-visually supplementing the narrative were not enough to garner the full appreciation of its audience.[112] Despite receiving very good-viewing figures, audience feedback was mixed and television critics roundly slated the documentary.[113] Although repeated the following year, aired across the world and supported by several different translations of the accompanying book, *The Life Game* never emulated the success of Calder's earlier documentaries.

The Life Game was quickly forgotten: its one-episode format, its initial poor performance and how soon the scientific content dated all contributing to its fleeting impact.[114] In stark contrast, *Ascent* guided by Bronowski's reassuring take on the evolutionary epic created an enduring footprint in British popular culture, continuing to be a source of interest for academics and popular audiences alike.[115] Like *Ascent*, *The Life Game* was an early humanist blockbuster, as both had big budgets generated via co-production agreements, and used multiple filming locations and cutting-edge graphics to celebrate humanity's scientific achievements. While *The Life Game* lacked a broader cosmological framework and the narrative arc of the evolutionary epic, its story about the ability of scientists to decipher nature's most complex games was an inherently progressive one. Where *Ascent*

for their own region, three, WNET, SR, and KRO/BRT, re-recorded their own studio sequences at the BBC. T63/69/1, BBC-WA.

[112] In recognition of his ability to communicate science clearly, Calder won the Kalinga Prize for the Popularization of Science in 1972. An award his father Ritchie had won just over a decade earlier in 1960.

[113] Letter from Philip Daly (Head of Science and Features, BBC) to Elisabeth de Vree (KRO), October 26, 1973, T63/69/1, BBC-WA; Peter Fiddick, "Television", *The Guardian*, October 15, 1973, 10; and Chris Dunkley, "Real Television", *The Financial Times*, October 17, 1973, 3.

[114] The series did find some more limited success where it was broadcast in a more educational specific context, rather than the Saturday evening slot it was given in the UK. For example, in Canada the documentary aired on the publically funded Ontario Educational Communications Authority, founded by the Education Minister in 1970 with a mandate to put educational programming onto commercial television channels.

[115] Unlike some of Calder's other documentaries, such as *The Violent Universe*, no recordings of *The Life Game* exist online. In contrast, *Ascent* was reissued on DVD in 2005, all episodes can be found in full on the Internet Archive, and it is still regularly referenced in both British popular culture and academic work. See: "History Documentaries", *Harry Hill's World of TV*, written and presented by Harry Hill, September 6, 2020, BBC2; and Zakariya, *A Final Story*, 278–287.

relied on the natural theological tradition of awe and spectacle, *The Life Game* focused on the majesty of recent biological discoveries and the minuscule probabilities involved in our existence to stoke wonder. Both celebrated human achievement and demonstrated that although the miracle of life connected all species, humans were exceptional.

It would be tempting, particularly for the narrative of this book, to account for the disparity in success between these two shows by simple appeal to their differing narrative forms. While *The Life Game's* "chaotic" chronology left it "lacking in a sense of direction", this was just one factor among several that contributed to its limited success. Again, as in earlier chapters, we must remember that individual scientists or science communicators did not create science broadcasts alone. The process of communicating science, especially via broadcast media, is a messy and contingent partnership between experts from multiple domains. While *The Life Game* was quickly forgotten, *Ascent* had a wide range of impacts beyond its immediate popularity. As well as influencing new television documentary formats, it also influenced the way that scientists themselves spoke about their work.

The Ascent of Man's *Influence on Metanarratives of Science*

Ascent continued to have a life well beyond its original airing, repeated many times in countries all over the world. In the US, as well as airing on public television in 1975, the series was widely used as an educational tool. By 1976, the series had been offered for college credit on history, anthropology and natural science courses in several hundred American universities.[116] In being used so extensively in educational settings, in the US and beyond, Bronowski's series became a key intergenerational vector in further coalescing and popularising the opinion that science was *the* progressive force of modern society.[117] In this regard, the series lived up to Bronowski's hope that by putting his synthetic totalising vision for science into a televisual, rather than written form, it would reach a much larger audience.

Bronowski was a key bridge figure, not just between scientists and popular audiences, but also between generations of synthetic thinkers. Bronowski's eldest daughter, the historian of science Lisa Jardine, has

[116] As well as university education, the US partners, Time-Life Inc. sold video copies and accompanying support materials to schools across the US. Marcel Chotkowski LaFollette, *Science on American Television: A History* (University of Chicago Press, 2013), 158–159.

[117] Zakariya, 285–287.

argued that C. P. Snow's infamous 1959 lecture on the two cultures was the culmination of a post-war debate about the role of science in British society.[118] I would argue that Bronowski extended this chronology. Just two years after *Two Cultures* and heavily influenced by Snow's reasoning, Bronowski's wrote his essay "Science is Human", which contained the main synthetic thrust later expounded in all its audio-visual glory in *Ascent*.[119] Thus, intellectually, *Ascent* should be considered as the long tail of a form of science communication that was still processing and rationalising scientific developments and their applications carried out under the auspices of the Second World War. The series was infused with Julian Huxley's approach of providing new visions, ideologies and routes forward for the secular humanist project, which both Huxley and Bronowski traced back to Renaissance Europe.

While intellectually, *Ascent* can be understood as the last popular endeavour of this group of British post-war intellectuals, it not only directly influenced undergraduates in the US but also a new generation of popular scientists. By the early 1970s, in academic forums at least, universal histories and synthetic narratives had fallen out of favour. Yet *Ascent* was to find an audience among a younger generation of scientists, who were interested in both synthetic thinkers, and the ability of popular mediums to bypass the slow, siloed processes of the academic world.[120] Up until his sudden death in August 1974, Bronowski continued to promote *Ascent* to popular and academic audiences.[121] *Ascent* found fertile ground, as several, mainly US-based, scientists had already begun experimenting with universal and mythic narratives, particularly in works for a popular audience. Tracing the influence of *Ascent* on later synthetic accounts—particularly those of astronomer Carl Sagan (1934–1996), theoretical physicist Steven Weinberg (1933–2021) and sociobiologist E. O. Wilson (b.1929)—Nasser Zakariya concludes that:

[118] Lisa Jardine, "C. P. Snow's Two Cultures Revisited", *The 2009 C.P. Snow Lecture*, given in Christ's College by Professor Lisa Jardine, October 14, 2009. As printed in *Christ's College Magazine* (2010), 49–57.

[119] Snow's lecture was published as a book the same year: C. P. Snow, *The Two Cultures and the Scientific Revolution* (Cambridge University Press, 1959).

[120] Zakariya, *A Final Story*, 275–278 later.

[121] For example in March 1973, *Ascent* in effect received its US premiere, when Bronowski used excerpts from the series in his keynote lecture at the annual meeting of the National Academy of Sciences in Washington. The National Academy of Sciences National Research Council, "Memorandum to Science Writers", March 27, 1973, T64/64/1, BBC-WA.

> [I]n wider intellectual exchange the series acted as the basis for discussion, reading Bronowski's visual and textual statement as a more firmly rooted representation of science and scholarship. ... If it did not last years in the future, that might have been as much or more the results of the success of the template it provided than a failure to resonate with what appeared to be an increasingly multiculturally minded audience. The next generation of public scientists overshadowed Bronowski's reception.

It is to the success of the template that *Ascent*, the archetypal humanist blockbuster, provided, that I now turn. Zakariya's account focuses on the individual scientists who fronted popular media, removing agency from the myriad other experts and external pressures that also shaped these popular works. As we have seen with *Ascent*, it was Adrian Malone and his production team who were responsible for the most important decisions that shaped the series' chronology, narrative form and its most "rhapsodic" audio-visual characteristics.[122] Humanist blockbusters not only share stylistic, format and production norms, but they also share a common lineage that can be traced via direct patronage, BBC departments, commissioning editors, producers and production teams back to Bronowski's *Ascent*.

Sledgehammers and Humanist Blockbusters, from *Ascent* to *Life on Earth*

Big-budget documentary series were in the ascendency during the 1970s, helping to establish television as a mature medium.[123] By the mid-1970s, the BBC had commissioned the "sledgehammer"[124] treatment on subjects as wide ranging as American history and economics.[125] As early as 1972, producers at the BBC Natural History Unit (NHU) in Bristol had proposed a major natural history series in the sledgehammer mould.[126]

[122] Zakariya, *A Final Story*, 280.

[123] Seaton, *Pinkoes and* Traitors, 122. For more on the earlier development of genres in scientific documentary see Boon, *Films of Fact*, Chapter 7.

[124] Attenborough, *Life on Air*, 214.

[125] *America*, written and narrated by Alistair Cooke, November 1972 to February 1973, BBC2; and *The Age of Uncertainty*, written and presented by Kenneth Galbraith, January–April 1977, BBC2.

[126] "Life on Earth", BBC Memo; and John Sparks, "Major Natural History Series", BBC Memo, August 8, 1972, WE 17/53/1, BBC-WA.

Alternative structures were proposed, most notably a 20-episode format fronted by two presenters, which could be easily edited for international markets. However, following contestation between the NHU and BBC2, the sledgehammer model prevailed, and what emerged followed *Ascent's* format directly; 13 parts with a presenter who was an integral component of the narrative arc of the series. During the production of *Life on Earth*, the producers regularly referred to *Ascent* as they looked to develop a style that was less superficial than leading US documentaries.[127]

David Attenborough, who had been a target for the natural history series since its earliest iterations, resigned from his role as Director of Programmes at the BBC in August 1972 to return to his career writing and presenting natural history broadcasts, primarily with the *Life on Earth* project in mind. Despite Attenborough's willingness, it was not until January 1974 when Aubrey Singer took over as the new controller of BBC2 that the project really took off. Building on his prior working relationship with the newly appointed Director of the NHU, Mick Rhodes (1935–2018), Singer made the series a top priority. By late 1975, Attenborough had drafted an outline of the series, and the producers Chris Parsons (1932–2002), Richard Brock (b.1938) and John Sparks (b.1939) had begun planning sequences.[128] As with prior series in this format, *Life on Earth* took several years to produce, filming on location in 49 countries. A co-production agreement between the NHU, Warner Brothers USA and Germany's Reiner Moritz Productions generated a final budget of over £1 million, at the time the most expensive TV series ever produced.[129]

Debuting on BBC2 on Tuesday January 16, 1979, *Life on Earth* began with the emergence of life, and explored the main evolutionary steps that led to the emergence of *Homo sapiens* (Table 6.4).[130] While *Ascent* clipped the chronology of the evolutionary epic to focus on the final seconds of man's scientific history, *Life on Earth*—although still lacking the deep

[127] "Life on Earth – Outline Proposal," BBC Proposal, 24 August 1972, T41/520/1, BBC-WA.

[128] For an in-depth account of the production and reception of *Life on Earth*, see: Gouyon, *BBC Wildlife Documentaries in the Age of Attenborough*, 205–244.

[129] For more on the scale and cost of the series see file SW3/32/1 and for a scrapbook of press clippings relating to the show from all over the world see WE 17/1, BBC-WA.

[130] Interested readers can find all of the original episodes of *Life on Earth* online at https://www.dailymotion.com/playlist/x6w54b *(last accessed 28/03/2021)*.

Table 6.4 Table listing the episodes and synopses for *Life on Earth*

	Episode	*Synopsis*
1	The Infinite Variety	David Attenborough travels back through time to find the earliest evidence of life on the planet and traces its development in the oceans of the world into a myriad of simple forms and ultimately the great reefs of coral. They were the first signs to be visible from space, indicating that life had started on Planet Earth.
2	Building Bodies	Larger creatures began to appear in the oceans. Some developed segmented bodies evolved an external armour and gave rise to crabs and lobsters. Others—such as clams and scallops—protected themselves with heavy stony shells. But a few, related to the nautilus, acquired bigger brains, abandoned their shells and relied for their survival on their great intelligence.
3	The First Forests	Some 400 million years ago, tiny moss-like plants together with segmented creatures such as millipedes began to emerge from the water to colonise the land. The plants, deprived of the support of water, had to develop woody stems to stand upright and reach the light. The animals had to find ways of transferring sex cells from one another without help of water—and that was a hazardous business.
4	The Swarming Hordes	A six-legged body with a hard covering proved to be a very adaptable pattern. It enabled insects not only to crawl and hop but to fly. Such creatures could not, however, grow very large, but some—the ants and the termites—dealt with that problem by forming colonies, each containing several million individuals and so became super-organisms that can dominate a plain or a jungle.
5	Conquest of the Waters	Backbones enabled fish to grow large and move at speed. Today they rule the seas, from the sun-lit world of coral reefs to the near-black freezing depths of the ocean where they communicate with luminous signals. Many undertake long annual journeys across the oceans. Some even swim up into freshwater to spawn—and there to die.
6	Invasion of the Land	Around 350 million years ago, some fish crawled up onto land. Their fins became legs; they breathed with the help of their wet skins and simple lungs. Amphibians—frogs, newts and salamanders—still live this way but their young initially need liquid in which to swim. Others keep their eggs moist in mounds of foamy spittle, in pouches on their backs or even within their throats.

(*continued*)

Table 6.4 (continued)

	Episode	*Synopsis*
7	Victors of the Dry Land	Reptiles developed a dry skin and a waterproof covering for their eggs. So equipped, they were able to colonise the driest lands. As they spread across the continents, so they evolved into a multitude of species. Dinosaurs dominated the world. Although they disappeared 65 million years ago, their relatives—snakes and lizards—still rule the hot deserts of the world.
8	Lords of the Air	Feathers are derived from the reptilian scale. Weight for weight they are stronger than any material mankind has invented. With feathers on their fore-limbs, birds took to the skies. But a coat of feathers brings other benefits. They keep a bird's body so warm that it can generate abundant energy; and they can acquire wonderful colours and shapes with which male birds perform courtship displays.
9	The Rise of the Mammals	Small four-legged furry animals scurried around the feet of later dinosaurs and when great reptiles suddenly died out, the descendants of these tiny creatures filled the vacancies. These were the first mammals. Two that survive—the platypus and the echidna—have warm furry bodies but still, like reptiles, lay eggs. Others, in the southern hemisphere, nurtured their young in pouches on their bellies.
10	Theme and Variations	In the northern hemisphere, mammals reproduced in a different way. The females retained their young within their bodies. At first, these mammals fed mostly on insects, as many still do—shrews and pangolins, giant ant-eaters and armadillos. Some—the bats—took to the skies in search of such food. Others returned to the seas from which their distant ancestors had come and evolved into whales and dolphins.
11	The Hunters and Hunted	When, some 30 million years ago, a change in the earth's climate caused the rain forests to shrink, they were replaced by grassy plains. Some plant-eating mammals ventured there to graze. But out in the open, they were vulnerable to hunting animals. So vegetarians became swift runners and gathered in herds for their mutual protection; and the big cats developed the weapons they needed to kill.
12	Life in the Trees	Some mammals took to climbing trees. To do that, they needed hands that could grasp and eyes that both looked forward to help in the judging of distance. The first to do so were creatures very like the bush-babies and lemurs of today. As time passed monkeys appeared. One group of their descendants became rather bigger and lost their tails. Their descendants included chimpanzees, gorillas—and ourselves.

(*continued*)

Table 6.4 (continued)

	Episode	*Synopsis*
13	The Compulsive Communicators	Grasping hands enabled human beings to use weapons. So they became hunters. But to kill the powerful plains-living animals, they needed to work in teams. That encouraged the development of language with which to share plans and issue instructions. Eventually, writing enabled them to share their experiences with others remote in both space and time. So human beings set off down the road towards the domination of the world.

Cover notes, *Life on Earth*, BBC DVD, (BBC Worldwide, 2005)

cosmological time of true evolutionary epics—could be considered *Ascent's* prequel.

Although favouring natural history location shoots over historical flashbacks, the opening episode still began with Darwin, the *Beagle* and the Galapagos Islands. Narrating a triumphalist account of Darwin's observations on speciation in the Galapagos, Attenborough then took the viewer into the fossil record and back to the earliest evidence for simple life forms on Earth. In stark contrast to Calder's *Life Game*, other than the Darwin framing used for the opening scene the process of scientific discovery was largely hidden throughout the series. It was not until the final episode, "The Compulsive Communicators" that *Homo sapiens* re-emerged. In earlier drafts of the series, there had been more time dedicated to humanity, however during several revisions of the script the production team dropped these scenes. The main contestations over the first draft script for the series were between the NHU and Attenborough. The NHU saw natural history film-making as a scientific process that contributed to biological knowledge, and as such they wanted the series to focus on new ideas and to go beyond accepted textbook knowledge. However, Attenborough saw the series more as an extension of *Zoo Quest*, and his role as scriptwriter needed only to draw on established secondary literature.[131] It was Attenborough's more glossy narrative-centred approach that won out, a key moment in his

[131] Highlighting the divergence in expertise and approach the NHU producer John Sparks, who was brought onto the project for his academic knowledge, had a PhD in zoology and had published on ornithology, while Attenborough had an undergraduate degree in natural sciences from 1947 and had limited experience outside of broadcasting. Mick Rhodes (Editor, NHU), Letter to Tom Williams, November 13, 1975, WE 17/56/1, BBC-WA; John Sparks and Tony Soper, *Penguins* (David & Charles, 1967) and John Sparks, *Bird Behaviour* (Hamlyn, 1969).

career carving out a new professional identity in wildlife filming; part writer, part producer and part presenter, what the historian of science Jean-Baptiste Gouyon has called the "telenaturalist".[132]

As with *Ascent* and earlier programmes in this format, the narrative arc of the series was of central concern. Senior executives were also concerned that Attenborough's style was too didactic. As Michael Peacock, Executive Vice President of the co-producers Warner Brothers, and former BBC2 Controller, confided in Mick Rhodes:

> Feature series such as … *Ascent of Man* were, in effect, collections of essays placed together within a chronological and intellectual scheme which made sense of the material by placing it in a landscape with plenty of familiar landmarks. … I sense that David [Attenborough]'s approach doesn't take sufficient account of these points, and in that sense is too didactic. My worry is that the diversity and beauty of life on earth which David brings out admirably is not seen against a broader perspective—the 'grand design' of nature, evolution or whatever—which fills in at least some of the how's and why's.[133]

Despite the machinations between Attenborough as scriptwriter and the NHU team filming the series, for those financing the project, if it was to appeal to a primetime international audience the narrative arc of the evolutionary epic had to be front and centre. In the discussions of the script, this tension remained central, with Chris Parsons and Mick Rhodes calling Attenborough's first draft "glossy further education". All three, influenced by Jacques Monod's *Chance and Necessity*, which Peacock had just read, pushed Attenborough to include more of the bigger frame, more scientific and philosophical connections and more of the difficult "how and why" questions.

Viewing figures for the series broke records for a documentary on BBC2, peaking at 15 million for episode nine. Furthermore, the

[132] Gouyon, *BBC Wildlife Documentaries in the Age of Attenborough*, 216–220; and Jean-Baptiste Gouyon, 'From Kearton to Attenborough: Fashioning the Telenaturalist's Identity', *History of Science* 49:1 (2011): 25–60, https://doi.org/10.1177/007327531104900102.

[133] Michael Peacock (Executive Vice President—Warner Bros TV), *Letter to Mick Rhodes* (Editor, BBC Natural History Unit), October 16, 1975, T41/520/1, BBC-WA.

audience and popular press response were exceptionally positive.[134] However, there were still some complaints along now-familiar lines. Of the 50–100 letters per day that Attenborough received during the broadcast run, a small number were complaints from "Fundamentalists" who took exception to the evolutionary storyline, in particular the connections drawn between apes and humans in the final episode. Anticipating a natural audience segmentation along these lines, when the repeat of *Life on Earth* went out on Sunday evenings, ITV re-aired the blockbuster religious drama *Jesus of Nazareth* (1977).[135] Using a rather more intellectual line of reasoning, the journalist Christopher Booker (1937–2019) heavily criticised *Life on Earth* in a series of essays in *The Spectator*, which focused on how a pseudo-mythic version of science had replaced prior Christian myths. Using familiar lines of attack, Booker criticised the neo-Darwinian consensus as unable to demonstrate that there is no organising principle behind evolutionary processes.[136] Booker raised the veneer of scientific authority afforded by the opening reference to Darwin, the lack of content on the unsolved mysteries of evolutionary science, and the polished mythic format of *Life on Earth*, to bolster his wider argument. Alongside the rehashing of disproven arguments, such as the lack of speciation in the fossil record, Booker also referenced biologists like Conrad Waddington (Chap. 4) who were unable to accept blind chance had resulted in humanity. A noted contrarian, it would be easy to dismiss Booker's essays as just anti-evolutionist sentiment cloaked in scientific language. However, among the inaccuracies, he made the astute observation that the series used the mythic narrative of the evolutionary epic to offer a simplified solution that could provide answers previously provided by traditional religion.

[134] Figures combined the estimated viewers for the original Tuesday broadcasts and the Sunday Repeat. Episodes received audience Reaction Indexes in the 83–90 range, by comparison, the regular NHU series *The World About Us* received an average RI of 76, and wildlife films in general received better RIs than other documentaries. "A review of press and viewer reaction", BBC Internal Report, January–April 1979, 23, WE 17/53/1, BBC-WA.

[135] "A review of press and viewer reaction", 7–8; Stephanie Cole, Letter to John Sparks, May 13, 1980, WE 17/53/1, BBC-WA.

[136] Christopher Booker, "What do you believe?", *The Spectator*, February 3, 1979, 16–17; Christopher Booker, "Darwin and Attenborough", *The Spectator*, February 10, 1979, 16–17; Christopher Booker, "In the Beginning ...", *The Spectator*, February 17, 1979, 14.

Life on Earth had a wide range of lasting impacts. At the BBC it completely realigned expectations with regard to natural history broadcasting, and along with other sledgehammers such as *Ascent*, it influenced a more expansive cinematic approach on regular science content like *Horizon*.[137] Alongside its widespread commercial success, many educationalists and other non-governmental organisations picked the series up. Widely used to gain support for conservation and environmental degradation issues, the series helped to popularise connections between the emergent discipline of ecology and our evolutionary history.[138] Attenborough's approach to "wildlife blockbusters" with their reliance on visual splendour and dramatic storyline is today ubiquitous (Chap. 1). The first steps towards this came during its immediate aftermath as the BBC producers responsible for *Ascent* and *Life on Earth* were in demand internationally and the format began to be replicated worldwide.[139]

Conclusion: The Humanist Blockbuster Goes Global

Many BBC staff originally involved with *The Ascent of Man* were instrumental to the success of subsequent humanist blockbusters, both at the BBC and beyond. As we have seen, Aubrey Singer and David Attenborough were both crucial to *Life on Earth's* success. Additionally, former BBC2 Controller, Michael Peacock in his new role at Warner Brothers, not only helped secure *Life on Earth's* funding but also shaped the narrative arc of the series. *Ascent* co-director and cameraman Mick Jackson worked on a number of US TV documentaries, including writing and directing an episode of *Nova*, a series itself modelled on the BBC's *Horizon*.[140] Most notably, *Ascent's* producer Adrian Malone, who was so instrumental in forming and refining the series' chronology and progressive narrative arc, was the executive producer on

[137] Brian James Kantor, 'Performing "the Authoritative Account": How the BBC's Horizon Produces Epistemic Authority', *Social Studies of Science*, 2020, https://doi.org/10.1177/0306312720947181, 5–6.

[138] For one rather apt example among several in the folder, see Kathy Bricker (Cedar Bough Films), letter to Dr [SIC] David Attenborough, April 7, 1982, WE 17/53/1, BBC-WA.

[139] Gouyon, *BBC Wildlife Documentaries in the Age of Attenborough*; Derek Bouse, *Wildlife Films* (University of Pennsylvania Press, 2011), Chapter 1; and Aldridge and Dingwall, "Teleology on Television?", 442–443.

[140] *Nova*. "Fusion: the Energy of Promise." Directed and written by Stuart Harris and Mick Jackson. PBS, May 19, 1974.

Carl Sagan's *Cosmos: A Personal Voyage* (PBS, 1980).[141] Sagan's hugely successful 13 part documentary series was the evolutionary epic writ-large, wide ranging in its scientific coverage it extended back beyond the scope of both *Ascent* and *Life on Earth* to the Big Bang itself.[142] In making Cosmos Malone replicated the "BBC paradigm", training the crew in the same house style, production and editing techniques central to his prior humanist blockbusters.[143] A year after Cosmos aired, in an interview with the *New York Times*, Malone described his approach: "I deal in the history and the making of ideas, and at the heart of it is synthesis". And thus, in training a US crew in this distinctly British mode of science broadcasting, with its progressive evolutionary narrative, Malone ensured the migration of this approach to a new generation of science broadcasters, from where it would go on to dominate globally.

This chapter began by considering how Bronowski first tried to popularise his synthetic scientific vision, his version of Huxley's scientific humanism, via the 1961 series *Insight*. While there were several reasons behind the series' failure, Bronowski's insistence on a thematic structure was the most telling. By the late 1960s when the BBC approached Bronowski regarding a new history of science series, the budget, technologies and importantly the format and distillation of the narrative form required to make a scientific epic a mainstream hit had progressed. The production team and in particular Adrian Malone were integral to pushing Bronowski's more esoteric outline into the format's confines. While other competing approaches to communicating evolutionary science, most notably *The Life Game* kept the progressive scientific lens, they lacked *Ascent's* splendour and mythic register. In explicitly discussing the contingent process of complex evolutionary science, they limited their audience and longevity.

[141] Incidentally, Carl Sagan had featured as the controversial scientist on an episode of *Controversy* (March 6, 1975, BBC2), the short-lived debate series introduced earlier in this chapter.

[142] For a detailed account of *Cosmos*, Adrian Malone's involvement and the direct links and comparisons with *Ascent* see Zakariya, *A Final Story*, 309–325. For more on *Cosmos* and its role in Sagan's media and scientific careers see Oliver Marsh, 'Life Cycle of a Star: Carl Sagan and the Circulation of Reputation', *The British Journal for the History of Science* 52, no. 3 (September 2019): 475–480.

[143] "The Ascent of Adrian Malone", *New York Times*, March 15, 1981, https://www.nytimes.com/1981/03/15/magazine/the-ascent-of-adrian-malone.html *(last accessed 21/06/2021)*.

By 1979 and *Life on Earth's* natural history take on the evolutionary epic, the sledgehammer format was a refined hit-making formula, already being exported to the US and beyond. While all of the major series discussed in this chapter focused on different aspects of the past, whether dealing with cosmic, natural or human history, they all had at their core progressive, secular, materialistic accounts of science and humanity. They were all humanist blockbusters.

References

Aldridge, Meryl, and Robert Dingwall, "Teleology on Television?: Implicit Models of Evolution in Broadcast Wildlife and Nature Programmes," *European Journal of Communication* 18:4 (2003): 435–453, https://doi.org/10.1177/0267323103184001.

Alland, Alexander. "Physical: *The Imperial Animal*. Lionel Tiger and Robin Fox." *American Anthropologist*, 75: 1147–1148. https://doi.org/10.1525/aa.1973.75.4.02a01600.

Attenborough, David. *Life on Air: Memoirs of a Broadcaster*. Princeton University Press, 2002.

Bada, Jeffrey L. and Antonio Lazcano, *Stanley L. Miller (1930–2007): A Biographical Memoir*. National Academy of Sciences (USA), http://www.nasonline.org/publications/biographical-memoirs/memoir-pdfs/miller-stanley.pdf.

BBC. *Annual Report and Accounts for the year 1958–1959*. HMSO, 1959.

———. *The British Broadcasting Corporation Handbook*. HMSO, 1968.

Beatty, John. 'The Creativity of Natural Selection? Part II: The Synthesis and Since'. *Journal of the History of Biology* 52:4 (2019): 705–731. https://doi.org/10.1007/s10739-019-09583-4.

Beaumont, Jen. "Population," *Social Trends 41*, Office for National Statistics Report. Crown Copyright, 2010.

"Best-Sellers," *Sunday Times*, 28 September 1975.

Black, Peter. "Peter Black's Teleview", *Daily Mail*, December 14, 1960.

Blundell, Tom. 'Orgel, Leslie Eleazer (1927–2007), Theoretical Chemist and Molecular Biologist'. Oxford Dictionary of National Biography. https://doi.org/10.1093/ref:odnb/99309.

Christopher Booker, "What do you believe?", *The Spectator*, February 3, 1979a, 16–17.

———. "Darwin and Attenborough", *The Spectator*, February 10, 1979b, 16–17.

———. "In the Beginning…", *The Spectator*, February 17, 1979c, 14.

Boon, Timothy. *Films of Fact: A History of Science in Documentary Films and Television*. Wallflower Press, 2008.

———. '"The Televising of Science Is a Process of Television": Establishing Horizon, 1962–1967'. *The British Journal for the History of Science* 48:1 (2015): 87–121. https://doi.org/10.1017/S0007087414000405.

Bouse, Derek. *Wildlife Films*. University of Pennsylvania Press, 2011.

Briggs, Asa. *The History of Broadcasting in the United Kingdom: Volume V: Competition*. Oxford University Press, 1995.

Bronowski, Jacob. *Insight*. Macdonald and Co., 1964.

———. *Magic, Science, and Civilization*. Columbia University Press, 1978.

———. *The Ascent of Man*. Little Brown, 1973a.

———. "Letters – 'The Ascent of Man'," *The Listener*, 23 August 1973b.

"Bryan Silcock Obituary", *The Times Online*, January 24, 2020, https://www.thetimes.co.uk/article/bryan-silcock-obituary-5ddhpz6vl.

Bucchi, Massimiano, and Brian Trench. *Handbook of Public Communication of Science and Technology*. Routledge, 2008.

Burke, Derek (Eds). *When Christians Disagree: Evolution*. IVP-UK, 1985.

Calder, Nigel. 'Science as Intellectual Entertainment'. *Nature* 245:5424 (1973): 293–295. https://doi.org/10.1038/245293a0.

———. *The Life Game; Evolution and the New Biology*. Viking Press, 1974. http://archive.org/details/lifegameevolutio0000cald.

Coady, Matthew. "Mirror TV: Your Complete Programme Guide," *The Mirror*, 21 May 1973, 18.

Cole, Rupert. '1972: The BBC's Controversy and the Politics of Audience Participation'. *Public Understanding of Science* 26:4 (2017): 514–518. https://doi.org/10.1177/0963662516684231.

Crick, Francis. *What Mad Pursuit*. Basic Books, 2008.

Crozier, Mary. "Television", *The Guardian*, January 4, 1961a, 5.

———. "Television", *The Guardian*, February 15, 1961b, 7.

———. "Television", *The Guardian*, April 5, 1961c, 5.

———. "Television," *The Tablet*, 2 June 1973, 13.

Cunningham, Andrew, and Perry Williams, "De-Centring the "big Picture": The Origins of Modern Science and the Modern Origins of Science," *The British Journal for the History of Science* 26:4 (1993): 407–432, https://doi.org/10.1017/S0007087400031447.

Desmarais, Ralph. 'Jacob Bronowski: A Humanist Intellectual for an Atomic Age, 1946–1956'. *The British Journal for the History of Science* 45:4 (2012): 573–589. https://doi.org/10.1017/S0007087412001069.

Dunkley, Chris. "The Ascent of Man," *The Financial Times*, May 7, 1973a.

———. "Real Television", *The Financial Times*, October 17, 1973b, 3.

———. "Blockbusters", *The Financial Times*, November 27, 1974, 3.

Edelson, Edward. *Gregor Mendel: And the Roots of Genetics*. Oxford University Press, USA, 2001.

Edgerton, David. "British scientific intellectuals and the relations of science, technology, and war," in *National Military Establishments and the Advancement of Science and Technology: Studies in 20th Century History*, ed. Paul Forman and José Manuel Sánchez Ron. Springer Science & Business Media, 1996.

Fiddick, Peter. "Television", *The Guardian*, October 15, 1973, 10.

Frantz Fanon, *The Wretched of the Earth*. Grove Press, 1963.

Gouyon, Jean-Baptiste. *BBC Wildlife Documentaries in the Age of Attenborough*. Springer Nature, 2019.

———. 'From Kearton to Attenborough: Fashioning the Telenaturalist's Identity'. *History of Science* 49:1 (2011): 25–60. https://doi.org/10.1177/007327531104900102.

Gregory, Jane, and Steven Miller. *Science In Public*. Basic Books, 2000.

Hall, Alexander. 'A Humanist Blockbuster? Jacob Bronowski and The Ascent of Man'. In *Rethinking History, Science, and Religion: An Exploration of Conflict and the Complexity Principle*, edited by Bernard Lightman. University of Pittsburgh Press, 2019.

Harman, Oren. 'Chance and Necessity Revisited'. *Journal of the History of Biology* 47:3 (2014): 479–493. https://doi.org/10.1007/s10739-014-9390-3.

Hearn, Marcus, "The Ascent of Man – Viewing Notes," BBC DVD/BBC Worldwide, 2005.

Herbert, Sandra. 'Darwin, Malthus, and Selection'. *Journal of the History of Biology* 4:1 (1971): 209–217. https://doi.org/10.1007/BF00356983.

Hesketh, Ian. "The Recurrence of the Evolutionary Epic," *Journal of the Philosophy of History* 9:2 (2015): 196–219, https://doi.org/10.1163/18722636-12341300.

Huxley, Julian. *The Humanist Frame*. Allen & Unwin, 1961.

Kantor, Brian James. 'Performing "the Authoritative Account": How the BBC's Horizon Produces Epistemic Authority'. *Social Studies of Science* (2020), https://doi.org/10.1177/0306312720947181.

LaFollette, Marcel Chotkowski. *Science on American Television: A History*. University of Chicago Press, 2013.

Latour, Bruno. *Science in Action: How to Follow Scientists and Engineers through Society*. Harvard University Press, 1987. http://archive.org/details/scienceinaction0000unse.

Leapman, Michael. "Singer, Aubrey Edward (1927–2007)", *Oxford Dictionary of National Biography*, Oxford University Press, 2011, http://www.oxforddnb.com/view/article/98830.

Lennon, Peter. "Man alive," *Sunday Times*, 6 May 1973, 38.

Lightman, Bernard. 'Darwin and the Popularization of Evolution'. *Notes and Records of the Royal Society* (2010). https://doi.org/10.1098/rsnr.2009.0007.

———. *Victorian Popularizers of Science: Designing Nature for New Audiences*. University of Chicago Press, 2009.

Marsh, Oliver. 'Life Cycle of a Star: Carl Sagan and the Circulation of Reputation'. *The British Journal for the History of Science* 52:3 (2019): 467–486. https://doi.org/10.1017/S0007087419000049.

McElheny, Victor K. "Jacob Bronowski Is Dead at 66; Leading Popularizer of Science", *The New York Times*, August 23, 1974, 32.

McGhee Jr., George R. 'Can Evolution Be Directional without Being Teleological?' *Studies in History and Philosophy of Science Part C: Studies in History and Philosophy of Biological and Biomedical Sciences*, Special Issue: Replaying the Tape of Life: Evolution and Historical Explanation, 58 (2016): 93–99. https://doi.org/10.1016/j.shpsc.2015.12.006.

Jacques Monod, *Chance and Necessity*. Penguin Books, 1997.

Jardine, Lisa. "C. P. Snow's Two Cultures Revisited", *The 2009 C.P. Snow Lecture*, October 14, 2009. *Christ's College Magazine* (2010), 49–57.

Morris, Bertram. 'Jacques Monod, Chance and Necessity: An Essay on the Natural Philosophy of Modern Biology (Book Review)'. *Technology and Culture; Detroit, Mich.* 13:4 (1972): 662–663.

"Plenty of time to grow up," *Sunday Times*, July 22, 1973, 52–53.

Radford, Tim. 'Calder, Nigel David McKail Ritchie (1931–2014), Journalist, Author, and Broadcaster'. Oxford Dictionary of National Biography, https://www.oxforddnb.com/view/10.1093/odnb/9780198614128.001.0001/odnb-9780198614128-e-108856.

Raj, Kapil. 'Beyond Postcolonialism ... and Postpositivism: Circulation and the Global History of Science'. *Isis* 104:2 (2013): 337–347. https://doi.org/10.1086/670951.

Ravetz, Jerome R. *Scientific Knowledge and Its Social Problems.* Transaction Publishers, 1973.

Richardson, Maurice. "Critic's Narrow Escape", *The Observer*, February 12, 1961, 31.

Robert Westman. 'The "Two-Cultures" Question and the Historiography of Science in the Early Decades of the Salk Institute for Biological Studies'. *Sartoniana* 32 (2019). http://www.sartonchair.ugent.be/file/293.

Rowan, John. "Letters – 'The Ascent of Man'," *The Listener*, 16 August 1973.

Russell, John. "Letters – 'The Ascent of Man'," *The Listener*, 21 June 1973.

Said, Edward. *Joseph Conrad and the Fiction of Autobiography*. Harvard University Press, 1966.

Seaton, Jean. *Pinkoes and Traitors: The BBC and the Nation, 1974–1987*. Profile Books, 2015.

Secord, James A. *Victorian Sensation: The Extraordinary Publication, Reception, and Secret Authorship of Vestiges of the Natural History of Creation.* University of Chicago Press, 2003.

Sparks, John. and Tony Soper, *Penguins*. David & Charles, 1967.

———. *Bird Behaviour.* Hamlyn, 1969.

Snow, C. P. *The Two Cultures and the Scientific Revolution*. Cambridge University Press, 1959.
"Television Today: Peter Willes wins RTS award," *The Stage*, 30 May 1974, 13.
"Television Today: Writers' Awards," *The Stage*, 16 May 1974, 13.
"The Ascent of Adrian Malone", *New York Times*, March 15, 1981.
Tiger, Lionel, and Robin Fox, *The Imperial Animal*. Dell Publishing, 1971.
Vorzimmer, Peter. 'Darwin, Malthus, and the Theory of Natural Selection'. *Journal of the History of Ideas* 30:4 (1969): 527–542. https://doi.org/10.2307/2708609.
Wagner, Roy. "Television Productions: The Ascent of Man," *American Anthropologist* 79 (1977): 993–994.
Wallace, Alfred Russel. *Travels on the Amazon and Rio* Negro. Ward Lock & Co., 1889.
Wicken, Jeffrey S. 'The Cosmic Breath: Reflections on the Thermodynamics of Creation'. *Zygon* 19:4 (1984): 487–505. https://doi.org/10.1111/j.1467-9744.1984.tb00943.x.
Wilson, Edward O. *On Human Nature*. Harvard University Press, 1978.
Wright, Pearce. "Gerald Leach", *The Guardian Online*, January 21, 2005, https://www.theguardian.com/science/2005/jan/21/obituaries.pressandpublishing.
Yates, F. Eugene, and Arthur S. Iberall. 'Chance and Necessity: An Essay on the Natural Philosophy of Modern Biology by Jacques Monod'. *Annals of Biomedical Engineering* 1:3 (1973): 381–384. https://doi.org/10.1007/BF02407677.
Young, Margaret Sery. "*The Imperial Animal*. by Lionel Tiger, Robin Fox", *Contemporary Sociology*, 1976, 5:3, 382–384. http://www.jstor.org/stable/2064160.
Zakariya, Nasser. *A Final Story: Science, Myth, and Beginnings*. University of Chicago Press, 2017.

CHAPTER 7

Creationism and Non-Darwinian Ideology in British Broadcasting

> The idea is that feathered birds arose by some gradual process from reptile ancestors, in which case there ought to be some intermediate links, but where are they? Is evolution just another creation myth? No-one has ever seen the birth of a major species or the change from one to another.[1]

In the spring of 1981, the BBC's long-running science documentary series *Horizon* aired an episode called "Did Darwin Get it Wrong?" The episode opened with the above quote, and posited that although controversial since 1859, Darwin's theory of evolution was now under attack from two camps: creationists and scientists themselves. In an early sequence, Duane T. Gish from the Institute for Creation Research in Texas set out the Creationist's literalist position on Darwinian evolution.[2] The dramatic opening built further, with commentary from noted anti-Darwinian critic

[1] *Horizon*, "Did Darwin get it Wrong?" Written and produced by Alec Nisbett, narrated by Paul Vaughan, March 30, 1981, BBC 2. The episode was repeated on BBC2 on April 5, 1981, and a version narrated by John Slack appeared in the US under the *Nova* brand via the PBS network. Interested readers can watch the US version of the documentary online at https://archive.org/details/DidDarwinGetItWrong *(last accessed 03/06/2021)*. All BBC copyright content reproduced courtesy of the British Broadcasting Corporation. All rights reserved.

[2] For more on Duane T. Gish and the *Institute for Creation Research* see: Ronald Numbers, *The Creationists: From Scientific Creationism to Intelligent Design*, expanded edition (Harvard University Press, 2006), 239–267.

A. Hall, *Evolution on British Television and Radio*, Palgrave Studies in Science and Popular Culture,
https://doi.org/10.1007/978-3-030-83043-4_7

and lawyer Norman Macbeth,[3] before the focus turned to the main content of the episode the latest findings in genetics. While new discoveries were challenging the neo-Darwinian status quo, they were ultimately still within the wider framework of evolutionary science.[4] Thus, despite the combative framing and press previews warning of "religious fundamentalists", the episode did not show alternative or new worldviews attempting to overrule the authority of Darwinian evolutionary theory.[5] Rather, the show presented a dichotomous framing that pitched science—both orthodox evolutionary views and new ideas challenging neo-Darwinian theory—against the religious, non-scientific and pseudo-scientific approach of US creationism. In effect, the show's producers used creationism—a relatively new, unknown and reportedly growing phenomenon—for its voyeuristic audience appeal.[6] By depicting creationism as something exotic, backward and dangerous, the documentary, which was really about how evolutionary thinking had developed since Darwin, used fear to generate interest and justify a position of scientific hegemony.

On reviewing the show in *The Times Literary Supplement*, columnist Eric Korn saw straight through this sensationalist artifice, criticising the "totemic power" of Darwin and arguing that we would not "intercut the latest upsetting findings in astrophysics with the doctrines of the flat earth

[3] Given Macbeth's background as a lawyer, many in the scientific community dismissed his argument for non-Darwinian evolution as put forward in his book *Darwin Retried: An Appeal to Reason* (Gambit Incorporated, 1971). The book focused on some of the most famous proponents of the Modern Synthesis, notably Ernest Mayr (1904–2005), George Simpson (1902–1984) and Julian Huxley, and attempted to take apart their popular works using legal techniques to highlight contradictions, but often conflated Darwinism and Neo-Darwinism and completely misunderstood the processes of scientific progress. See: John Maynard Smith, "Counsel for the defence," New Scientist, 22 August 1974, 476.

[4] The episode featured the following scientific ideas and evolutionary concepts: point mutation, jumping genes, gaps between genetic level theories and organism level evolution, bifurcation or branching of characteristics and neo-Lamarckian inheritance. In total, the one-hour broadcast only gave creationist and other non-scientific critiques of evolution around 7.5 minutes of airtime.

[5] Cowley, E. "Pick of the Day: Horizon." *The Daily Mail*, March 30, 1981; and "Personal Choice." *The Times*, March 30, 1981.

[6] The show was written and produced by long-serving *Horizon* producer Alec Nisbett, with input from series editor Simon Campbell-Jones; short interviews with both, in which they discuss the editorial and decision-making process of creating *Horizon* episodes can be watched online: "Horizon at 50-interviews", *History of the BBC*, https://www.bbc.com/historyofthebbc/research/horizon-at-50-interviews *(last accessed 05/06/21)*.

society, as though the one supported the other".[7] Given that Korn was the only reviewer who mentioned the creationist element, and the majority of reviews said the show was boring, it is probably safe to say that the producer's attempts to heighten drama through this dichotomous framing were in this instance, largely unsuccessful. The frame adopted for this episode of *Horizon* was part of a wider approach at the BBC, which as per their charter, attempted to provide balance and give space for opposing views. This approach, particularly with regard to science, has regularly been criticised for giving a false equivalence to contradictory views, especially those which are not supported by empirical evidence.[8]

In 1981, following "Did Darwin Get it Wrong?", the BBC went on to air another five radio and television broadcasts about creationism. However, before we get to the spike in coverage on creationism in the British media during the 1980s, we must first revisit some of the earlier coverage of anti-evolution positions at the BBC. We have already seen how in the 1940s and 1950s the BBC dealt with co-ordinated anti-evolution campaigns from fundamentalist religious groups, and how the frequency of these letters led Archibald Clow to produce a Report on Evolution in 1952 (Chaps. 3 and 4). As in this earlier period, the subsequent decades saw such positions given almost no serious airtime on British broadcast media. All non-scientific criticism of evolution was relegated to the fringe of the schedule, and only occasionally fleetingly featured in programmes, such as radio dramas and philosophical lectures. This chapter first introduces the few sporadic occasions when non-scientific anti-evolution critiques received airtime on the BBC, before covering the marginally more frequent instances when scientific content—albeit that covering fringe and pseudo-scientific ideas—challenged neo-Darwinian consensus. The chapter then returns to the coverage of creationism in the 1980s, highlighting that while such content inadvertently gave a platform

[7] Eric Korn, "remainders", *The Times Literary Supplement*, May 15, 1981, 541.

[8] While today readers may be most familiar with this criticism in relation to the BBC's coverage of the climate crisis (e.g. Bob Ward, "The BBC is sacrificing objectivity for impartiality in its coverage of climate change", *British Politics and Policy at LSE*, 2012. http://eprints.lse.ac.uk/id/eprint/48416), it is a charge that has long been raised at science programming generally, and the show *Horizon* specifically. For example, the *Horizon* episode, "Worlds in Collison" (1973), which dedicated an entire hour to the discredited pseudo-scientific and pseudohistorical theories of Dr Immanuel Velikovsky (1895–1979) was criticised by scientific commentators. For some examples of the reaction to this episode of *Horizon* see: "Letters", *New Scientist*, January 25, 1973, 210–211.

to anti-evolutionist ideology, in actuality it mainly used creationism as a hook to draw in an audience. Ultimately, this programming, whether made by science producers or the religion department, sought to other anti-evolutionist ideas and defend the authority of evolutionary science.

Limited Airtime for Fundamentalists and Anti-evolutionists

Prior to 1981, the only British broadcasts that featured creationism were on the 1925 Scopes Trial, in which American high-school teacher John Scopes contested Tennessee's Butler Act (1925), which prevented the teaching of human evolution from lower orders of animals in any public school in the state.[9] The first of these broadcasts was a radio adaptation of the play *Inherit the Wind*, first aired in late 1965.[10] The same year, ITV broadcast the film adaptation of the play starring Spencer Tracey as fictionalised defence lawyer Henry Drummond.[11] Although it featured some direct excerpts from courtroom transcripts, *Inherit the Wind* was a fictionalised account of the trial, used as a parable to criticise the assault on intellectual freedom then underway in American society under the guise of the McCarthy era trials.[12] While the film was critical of creationism, it did not present a simple dichotomous conflict between science and religion.[13]

[9] For more on the history of the Scopes Trial, and in particular the divergence from popular accounts which frame it as primarily about science versus religion, rather than as a contestation between religious modernists and fundamentalists in the US, see: Adam R. Shapiro, *Trying Biology: The Scopes Trial, Textbooks, and the Antievolution Movement in American Schools* (University of Chicago Press, 2013), 1–13.

[10] *Inherit the Wind*, written by Jerome Lawrence and Robert E. Lee, adapted by Nicholas Bethell, BBC Home Service, December 27, 1965. Lawrence and Lee's play premiered in Dallas in January 1955, and following rave reviews ran on Broadway from April 1955 until 1957. The only other BBC broadcast on the Scopes Trial pre-1980s was part of a BBC Radio 4 series on famous US trials broadcast in 1967–1968: "Evolution in Tennessee", *Famous American Trials*, written and narrated by Edgar Lustgarten, August 14, 1968, BBC Radio 4.

[11] The playwrights called the defence lawyer Henry Drummond in a knowing nod to the 19th evangelical and theistic evolutionist Henry Drummond (1851–1897). Bernard Lightman, 'Darwin and the Popularization of Evolution', *Notes and Records of the Royal Society*, 2010, https://doi.org/10.1098/rsnr.2009.0007; and "today's tv and sound," *The Observer*, October 10, 1965, 22.

[12] "Inherit the Controversy", *Newsday* (New York), March 17, 1996, 85.

[13] Despite the historical context of the plays' publication and the authors' original intentions, *Inherit the Wind* has not always been interpreted as fiction, particularly in the US where the play and film have remained popular in educational contexts. See Randy Moore,

The very first mention of *Inherit the Wind* on British television had actually been several years earlier, in an October 1960 episode of the Religion Department series *Viewpoint*, titled "Can We Bury the Hatchet?"[14] The episode opened with a courtroom clip from the film adaptation, which the writer and presenter John Wren-Lewis (1923–2006) used to frame his argument that Christians who opposed Darwin were wrong. This was not for the usual fundamentalist's reasons, he argued, but rather because Darwin, like Genesis, removed agency from nature and thus more primitive forms of spirituality:

> This is why the Church should have welcomed him if it had been about its business—not because he showed how God created nature, but because he made it practically impossible to believe in a God behind nature, and so struck a telling blow at the most persistent kind of superstition known to man.[15]

Wren-Lewis was a research executive in the British chemical industry who developed an interest in science and religion issues, and in the early 1960s was a central figure in the Death of God movement. Alongside his popular writing on science, education and religion he was still working in the chemical industry in this period, and he later became a gigging academic lecturing and writing on various aspects of humanistic psychology.[16] Wren-Lewis sought to draw a distinction between Darwin's original science and the subsequent philosophy of evolution, which would make "people regard human life as a means rather than an absolute".[17] While Wren-Lewis' viewpoint in this episode was somewhat eccentric, in keeping with earlier BBC programming on evolution and religion, such as *Science and the Christian Man* (Chap. 3), it was still ultimately an accommodationist argument, which did not question the hegemonic authority of science.

'Creationism in the United States: VII. The Lingering Impact of "Inherit the Wind"', *The American Biology Teacher* 61:4 (1999): 246–250, https://doi.org/10.2307/4450666.

[14] *Viewpoint* was a fortnightly programme on BBC TV, which aimed to give a Christian perspective on current affairs and topical problems. A flagship series for the Religion Department it ran from 1959 to 1973.

[15] "Can We Bury the Hatchet?", *Viewpoint*, written and presented by John Wren-Lewis, produced by Vernon Sproxton, October 5, 1960, BBC TV, microfilm transcript, BBC-WA, 6.

[16] For a collection of essays and articles written by John Wren-Lewis visit: https://www.capacitie.org/wren/archive.htm *(last accessed 10/06/21)*.

[17] "Can We Bury the Hatchet?" microfilm transcript, BBC-WA, 7.

Elsewhere, where broadcasters brought religious and philosophical perspectives into conversation with evolutionary science, this accommodationist framing remained front and centre. On the rare occasion when content did transgress a neat accommodationist relationship between evolution and religion, it never came at the expense of scientific authority; in most instances, it actually proposed further integration between the two institutions, or envisioned future directions for humanity and our belief systems as informed by new scientific understandings. Take for example the 1967 BBC Radio 4 *Reith Lectures*; given by the social anthropologist Edmund Leach (1910–1989) titled "A Runaway World?"[18] In this provocative series of six, 30-minute lectures, Leach proposed that as science had enabled humanity to become God-like, society should seriously explore its own divinity.[19] Influenced by Julian Huxley, and with echoes of his episode in the *Reshaping Man's Heritage* (1943) series (Chap. 2), Leach argued that we should not fear our mastery of nature and that scientists should play an active role in shaping our evolution.[20] Throughout, he framed a purported lacuna in scientists' exploratory approach through an explicitly evolutionary-humanist lens:

> Darwin and his friends were thought to be dangerous atheists, but their heresy simply replaced an inherent impersonal deity called God by a benevolent personal deity called Evolution. In their different ways Bishop Wilberforce and T. H. Huxley both believed in fate. It is this religious attitude which still dominates all scientific thinking about future development.[21]

Like Bronowski's account in *The Ascent of Man* just a few years later, Leach's vision relied on a version of science that was inherently neutral, with problems only emerging when individuals misapplied scientific

[18] *The Reith Lectures*, "A Runaway World?", written and presented by Edmund Leach, November to December 1967, BBC Radio 4. The lectures were also repeated in an earlier slot the following day on BBC Radio 3, and serialised in the BBC magazine *The Listener*.

[19] The six talks were titled, "Men and Nature", "Men and Machines", "Ourselves and Others", "Men and Morality", "Men and Learning" and "Only Connect". Recordings of parts three and six of the lectures can be listened to online: https://www.bbc.co.uk/programmes/p00h3xy8/episodes/player *(last accessed 10/06/21)*.

[20] When drafting the lectures Leach took as a starting point the concluding remarks of a 1963 conference on "Man and His Future" between Julian Huxley, Peter Medawar, and neurologist Russell Brain (1895–1966). Edmund Leach, *A Runaway World? The B.B.C. Reith Lectures 1967* (Oxford University Press, 1968), vii–ix.

[21] Edmund Leach, *A Runaway World?*, 5.

knowledge. However, Leach's lectures were light on the detail of how humanity might best manage such co-ordinated scientific progress, leaving a lot of the future heavy lifting to "the young".[22] As with *Ascent*, Leach's approach—particularly his reliance on universal solutions and his techno-optimism—was out of kilter with nascent counter-cultural movements of the period, especially growing awareness of the environmental degradation caused by industrialised societies. The editor of progressive Catholic newspaper, *The Tablet*, noted the similarity in appeal between Leach's humanist progressive vision and the popular views of French palaeontologist, Jesuit priest and vitalist, Teilhard de Chardin (1881–1955).[23] Yet to modern listeners, many of Leach's arguments—such as the call to collapse the distinction between other animals and humans, and between humans and machines—would seem prescient. Unsurprisingly, his lectures provoked an angered response from several sections of British society.[24] Even those audience members who found Leach's lectures enjoyable and thought-provoking found parts alarming. As one respondent to the BBC audience research survey summarised: "'Scientists as Gods'—it had to be said sooner or later, at least it was said with humility".[25] While others complained that the speaker was too left wing and that the central metaphor was philosophically illogical, the majority of objections were on religious grounds, particularly the narrow materialist focus of Leach's worldview.[26]

Examples, like Leach's *Reith Lectures*, when alternatives to the status quo of popular British discourse were aired whether on evolutionary science or religious worldviews, are limited and most often found in similar highbrow content, which was speculative and deliberately provocative. In

[22] In a follow-up debate on BBC Radio 3 featuring Leach, Peter Medawar and Alasdair MacIntyre (1929–present), this question of how best to manage such a scientific co-operative future was discussed in more depth. "A Runaway World Pursued", January 1, 1968, BBC Radio 3. For more on this debate and Leach's Reith Lectures see: Stanley J. Tambiah, *Edmund Leach: An Anthropological Life* (Cambridge University Press, 2002), 380–403.

[23] Douglas Woodruff, "Talking at Random", *The Tablet*, December 2, 1967, 23.

[24] Edmund Leach, *A Runaway World?*, 18.

[25] BBC Audience Research Report, "The Reith Lectures 1967: A Runaway World? by Edmund Leach, 1: Men and Women [SIC]", December 29, 1967, BBC-WA.

[26] Reflecting the huge social changes that had occurred in British society since Huxley's early scientific humanist broadcasts, the BBC, and in particular the Head of Talks, George Camacho staunchly defended Leach's selection, and their remit—as had been highlighted in the Pilkington Committee Report of 1962, to give reasonable airtime to humanist and agnostic views. See R51-1252/3, BBC-WA.

relation to creationist arguments, one pre-1980s broadcast stands out: an episode of the Religion Department radio series *The People and the Book* (1975) titled "Leave my Ancestors in Paradise". The series, presented by journalist and broadcaster Brian Redhead (1929–1994), looked back at history to see how contemporary understanding of the bible, particularly in light of growing secularisation in the UK, had emerged.[27] Quoting a nineteenth-century response to Darwin's *Origin*, the fifth episode began, "[l]eave my ancestors in Paradise, and I will leave yours in the Zoological Gardens". It then proceeded to provide a selective history of the conflict between evolutionary science and Christian theology, outlining how the schism between science and religion had grown across the first half of the nineteenth century. Like Orwell's episode in the *Voyages of Discovery* (1946) series (Chap. 5), the episode's sole historical focus centred on nineteenth-century conflict between evolutionary science and traditional religious beliefs:

> Wilberforce and people who thought like him were right; there was no room for agreement between Darwin and Christianity if their view of Christianity was the right one. If the plan of salvation really did depend on the literal truth of the Book of Genesis, then science was indeed the enemy of religion, because science was coming up with more convincing explanations of physical facts than any that Genesis had to offer. Darwin had written a scientific study—but it was attacked by Churchmen as if it were a theological treatise. And perhaps this was inevitable, because for generations, churchmen had been treating Genesis as if it were a scientific study. That was an attitude that could only lead to conflict.

Similar to most science programming in the period (Chap. 8), the episode used this oversimplified conflict-centred narrative of Victorian Britain's response to Darwinism as a counterfoil to the more enlightened position apparently popular in contemporary British society.[28] Concluding:

> Because these false antagonisms have been infinitely dangerous to faith, over Darwin's grave, let us once more assure the students of science that we

[27] *The People and the Book*, presented by Brian Redhead, written and produced by Fraser Steel, September to December 1975, BBC Radio 4.

[28] For a recent history that deals in more detail with the nuance of late nineteenth-century science and religion debates triggered by Darwin and Wallace's publications see: James C. Ungureanu, Science, Religion, and the Protestant Tradition: Retracing the Origins of Conflict (University of Pittsburgh Press, 2019).

> desire the light. We believe in the light. We press forward into the light. If need be, let us perish in the light. But we know that in the light we shall never perish. For to us, God is light; and Christ is, and will be to the end, the light of the world.[29]

This narrative device of using a case study that was external to listeners own experience, whether historically or geographically, allowed those listening at home to feel enlightened and perhaps slightly superior. This device was repeated on content focused on US creationism in the 1980s. However, before we return to these broadcasts and their interrogation of the associated pseudo-scientific Intelligent Design movement, we first need to know how non-mainstream evolutionary ideas, particularly those attempting to contest Darwinian and neo-Darwinian ideas on scientific grounds, were covered in the preceding decades.

Non-Darwinian and Post-Darwinian Evolutionary Ideas

Given how much knowledge of evolutionary science has changed since the first broadcasts on evolution in 1925, it is inevitable that some of the scientific ideas and concepts broadcast during the following decades subsequently proved incorrect. However, during the post-war decades there were sporadic instances when broadcasts featured fringe scientific positions that were controversial at the time. Some went on to become scientific consensus, such as Motoo Kimura's neutral theory of molecular evolution featured in *The Life Game* (1973), but many were categorically disproven or have remained in the pseudo-scientific wings of popular culture. More often than not, fringe scientific views, featured in broadcasts as counter-points to the mainstream consensus, used to heighten dramatic tension in a documentary, or to meet the requirement for the public service broadcaster to present a balanced view.

Unsurprisingly, scientists who have maintained a public profile through writing popular books have also received a disproportionate amount of airtime. As myriad media studies have shown, controversy and

[29] "Leave me my Ancestors in Paradise", *The People and the Book*, presented by Brian Redhead, written and produced by Fraser Steel, October 19, 1975, BBC Radio 4. Microfilm, Talks Scripts 167/168, BBC-WA.

dramatisation often ensures media coverage and higher audience figures.[30] Occasionally, we can see a form of generational inertia whereby established scientists with a large public profile continued to be given a popular platform, despite no longer producing leading research on the subject matter. For example, the zoologist and broadcaster Desmond Morris, whose controversial popular book *The Naked Ape* (1967), which was in large part sociobiological conjecture and anecdotal in its approach, featured regularly across the BBC.[31] This included a notable 1977 appearance on the chat show *Parkinson*, where following an argument from *The Naked Ape* Morris insinuated that the fullness of actor Diana Dor's lips might have contributed to her success.[32]

[30] For discussion of this effect in relation to science journalism see: Maxwell T. Boykoff, *Who Speaks for the Climate?: Making Sense of Media Reporting on Climate Change* (Cambridge University Press, 2011), 100–109; Timothy Caulfield, 'Popular Media, Biotechnology, and the "Cycle of Hype"', The Mass Media's Influence on Health Law and Policy Symposium, *Houston Journal of Health Law & Policy* 5:2 (2005): 213–234.

[31] From 1955 onwards Desmond Morris was a regular on BBC natural history content, particularly children's educational radio, and as Head of the Granada TV and Film Unit at the Zoological Society of London, he was a regular on ITV, particularly through their flagship natural history programme *Zootime* (Chap. 4). Despite moving to a curatorial role at the Zoological Society of London in 1959, he remained a regular on a wide range of British broadcasts through the 1960s. *The Naked Ape* received widespread media attention, most notably, it was serialised in the *Daily Mirror* and was the subject of a dedicated episode of the natural history series *Life* (1965–1968) on which Morris was a regular contributor and narrator. Although well received in the popular press, Morris' *Naked Ape* was widely criticised by academics, see for example: Russell H. Tuttle, review of 'Review of The Naked Ape: A Zoologist's Study of the Human Animal; The Apes: The Gorilla, Chimpanzee, Orangutan and Gibbon – Their History and Their World, by Desmond Morris and Vernon Reynolds', *American Anthropologist* 70:6 (1968): 1238–1240; and John Lewis and Bernard Towers, *Naked Ape or Homo Sapiens?* (Garnstone Press, 1972), http://archive.org/details/nakedapeorhomosa00lewi. More recently, the book has been criticised for its reductionist overreliance on the mechanism of natural selection, and as an example of poor science communication, blurring the line between speculative conjecture and empirical scientific consensus, which helps embed erroneous scientific ideas in popular culture. See: Nico M. van Straalen, 'The Naked Ape as an Evolutionary Model, 50 Years Later', *Animal Biology* 68:3 (2018): 227–246, https://doi.org/10.1163/15707563-17000167; Adam Rutherford, *Humanimal: How Homo Sapiens Became Nature's Most Paradoxical Creature—A New Evolutionary History* (The Experiment, 2019); Erika Lorraine Milam, *Creatures of Cain: The Hunt for Human Nature in Cold War America* (Princeton University Press, 2019), 113–123.

[32] *Parkinson*, October 8, 1977, BBC One. A short clip of the exchange between Morris and Dors can be found online at https://www.youtube.com/watch?v=tvVa0bpeeGM *(last accessed 15/06/21).*

Another zoologist with a popular media profile in the period was Alister Hardy (1896–1985), who's pseudo-scientific aquatic ape hypothesis, first published in 1960, briefly received the attention of popular science media, including at the BBC.[33] The theory, largely ignored by professional scientists at the time, attempted to explain away some of the mysteries of human evolution—like why unlike other primates we have a layer of subcutaneous fat—by positing that our ancestors spent a period of time evolving in a semi-aquatic landscape. Reflecting the speculative nature of his original thesis, even Hardy himself kept it separate from his other popular writing. For example, there was no mention of man's aquatic ancestry in *The Living Stream* (1965), based on his 1963 Gifford Lectures, which incidentally drew heavily on Julian Huxley's "stream of life" metaphor, as first popularised in his 1925 BBC lecture series.[34] Even Desmond Morris, a former student of Hardy's, deemed the "ingenious" theory lacking in enough direct evidence to be given serious consideration.[35] Moreover, although today the theory in its original manifestation still lacks any major direct substantiating evidence, it continues to sporadically reappear in popular culture; its amorphous boundaries helping to ensure its ongoing appeal.[36] The theory most recently re-emerged into the popular conscience following a two-part radio broadcast by David Attenborough, *The Waterside Ape*, which aired on BBC radio in September 2016.[37] The broadcasts triggered a swift and strong rebuttal by some in the scientific community. In turn, another group of scientists followed this rebuttal by

[33] See Alister Hardy, "Was Man More Aquatic in the Past?", *The New Scientist*, March 17, 1960, 642–645; and *Has Man an Aquatic Past?*, written and presented by Alister Hardy, April 29, 1960, BBC Third Programme.

[34] Alister Hardy, *The Living Stream* (Harper & Row, 1965), 30–40.

[35] Desmond Morris, *The Naked Ape: A zoologist's study of the human animal* (Jonathan Cape, 1967), 40–45.

[36] Robert Foley and Marta Mirazón Lahr, 'The Role of "the Aquatic" in Human Evolution: Constraining the Aquatic Ape Hypothesis', *Evolutionary Anthropology: Issues, News, and Reviews* 23:2 (2014): 56–59, https://doi.org/10.1002/evan.21405. For a historical account of one of the most interesting re-workings of Hardy's original thesis, writer Elaine Morgan's feminist deconstruction of human anthropologies overly male-centred focus as first outlined in *The Descent of Woman* (1972), see Erika Lorraine Milam, "Dunking the Tarzanists: Elaine Morgan and the Aquatic Ape Theory", in Oren Harman and Michael R. Dietrich (Eds), *Outsider Scientists: Routes to Innovation in Biology* (University of Chicago Press, 2013), 223–234.

[37] *The Waterside Ape*, written and presented by David Attenborough, September 14 & 15, 2016, BBC Radio 4. Available online at https://www.bbc.co.uk/programmes/b07w4y98/episodes/player *(last accessed 16/06/21)*.

publicly supporting Attenborough for giving airtime to cutting-edge new discoveries.[38] For our purpose, there is no need to dwell on which group is correct. Rather, this contestation highlights an interesting point about scientific expertise and cultural authority in the mass media. What was such a respected broadcaster doing in this instance that so incensed some prominent scientists, and why did the majority of the public who commented support the broadcaster? By limiting the function of popular science media to being just a vehicle for the dissemination of established scientific knowledge, the scientists who criticised David Attenborough's broadcasts misjudged how human persuasion works and how storytelling devices work to reinforce or enhance positions of cultural authority. By focusing purely on the scientific content of Attenborough's message—the logos—the scientists criticising him forgot to account for his utilisation of pathos and the currency and trust—the ethos—that he had built up with his audience over the last 60 years of his broadcasting career. If we wish to democratise science, demanding of citizens that in addition to knowing scientific facts, they must also understand scientific process, then as with the *Waterside Ape* and *Controversy* in the 1970s (Chap. 6), even factual science broadcasts must be a space where discussion, disagreement and new ideas can be aired. It is to more recognisably creationist ideas, and their coverage by British broadcast media in the 1980s to which we now return.

Reporting on US Affairs or Creating Creationists? Creationism at the BBC

Despite the instances introduced when fringe scientific and pseudo-scientific ideas received airtime, prior to the 1980s BBC science broadcasts had never tolerated "scientific" versions of creationism as popular in the US. As Nigel Calder surmised in 1973:

> What should we tell Charles Darwin, if he rejoined us after ninety-odd years in the grave? He would need to learn about the survival of the quickest,

[38] See Alice Roberts and Mark Maslin, "Sorry David Attenborough, we didn't evolve from 'aquatic apes'—here's why", *The Conversation*, September 16, 2016, https://theconversation.com/sorry-david-attenborough-we-didnt-evolve-from-aquatic-apes-heres-why-65570 *(last accessed 17/06/2021)*. *The Guardian*, *The Independent* and the *Scientific American* among others picked up this online article.

> when it came to crossing the street, and he would be glad, of course, that men and women are still interested in that question of questions—how do we come to be here? Learning that the state of California is now editing its school biology books so as to avoid offence to the Bible thumpers might irritate him, but probably no more than overhearing the remark, "Darwin said it all".[39]

Following Calder's broadcast in 1973 (Chap. 6), the cultural battles being fought in the US between fundamentalists and progressives, particularly those centred on the teaching of evolution in classrooms continued to escalate. By the early 1980s, popular support for US Creationists' organised pseudo-scientific front, Intelligent Design, had grown enough to pique the interest of journalists in the UK. The *Horizon* episode, "Did Darwin Get it Wrong?", was the first of a series of broadcasts (Table 7.1) that introduced the British public to creationist ideologies including Young and Old Earth Creationism, Scientific Creationism and Intelligent Design.

In covering these US phenomena, the BBC rarely distinguished between the different camps and ideologies, most commonly referring to creationists as a nebulous whole, and connecting the movements with American Evangelical Christianity and US far-right political groups.[40]

In fact, most of the approximately 30 programmes on creationism that the BBC have aired since 1981 (Table 7.1) were broadcast in the 1980s, the decade that saw creationism remerge as a mainstream issue in the United States.[41] The uptick in programmes was part of wider British media interest in covering the US movement. Creationism, and, specifically, scientific creationism, made for popular content amongst British audiences.[42] Here we see how mass media acts as a reactive device reflecting societal

[39] Nigel Calder, *The Life Game: Evolution and the New Biology* (Viking Press, 1974), http://archive.org/details/lifegameevolutio0000cald, 9.

[40] For example, *God-in-a-Box*, written and presented by Colin Morris, January 10, 1984, BBC Two; and the subsequent book that followed the programme: Colin Morris, *God-in-a-box: Christian strategy in the television age* (Hodder and Stoughton, 1984), 194–199.

[41] Numbers, *The Creationists*, 351–362.

[42] The Google Book N-Gram Viewer for the words "creationist" and "creationism" shows, for the American English Corpus (1800–2009) a sharp uptick beginning in 1975, and for the British English Corpus (1800–2009), a slightly smaller and delayed, uptick in the words' use beginning in around 1979. In both corpora frequency peaks in around 1985–1986, and again from a second wave in around 2007–2008. A similar pattern can be observed when running term frequency enquiries on UK newspaper databases, such as the Gale Primary

trends, which by focusing on particular narratives, in turn itself shapes, influences and amplifies popular discourse on the subject.

A wide range of BBC departments, including Features, Drama, Talks, Education and Religion produced broadcasts on creationism (Table 7.1). They include an episode of daytime radio chat show *Woman's Hour*, on which, in their regular feature "Science Simplified" they asked, "Is evolution a fact, or an elaborate fiction?"[43] Two of the shows were part of the long-running Religion Department series *Everyman* (1977–2005), which explored moral and religious issues in society, while three broadcasts were part of the long-running science series *Horizon*. All were categorical in their outright rejection of scientific creationism. While the exact narrative approach and framing of the issues varied greatly depending on which department made the content, all of these broadcasts promoted a normative position that science and religion, at least in some contexts, were in conflict. Five of the programmes in the table were repeats of the 1965 film *Inherit the Wind*, and another seven were broadcasts, or repeats, that focussed on the 1925 Scopes Monkey Trial. As with other Open University content introduced in Chap. 4, we see the content made under the BBC/OU partnership regularly repeated for different cohorts of students. The television programme *The Tennessee Evolution Trial* (1981) was largely researched and written by OU historian of science James Moore and, as such, was a much more historically contextualised and nuanced treatment of the Scopes trial than shows produced by BBC science broadcasters.[44]

As with the Darwin flashbacks introduced in Chap. 5 and picked up again in Chap. 8, the majority of the remaining shows in Table 7.1 used the 1925 Scopes trial as an opening contextual device. A foundational

Sources collection. For another broadcast example see: "87—Thy Kingdom Come", *Viewpoint*, written and presented by Anthony Thomas, April 14, 1987, ITV.

[43] *Woman's Hour*, presented by Sue MacGregor, July 30, 1981, BBC Radio 4. Partial transcript available on microfilm at BBC-WA.

[44] The broadcast was part of the course materials for the module, 'Science and belief: from Darwin to Einstein' (A381), which ran at the Open University from 1981 to 1987, and followed on from the course 'Science and belief: from Copernicus to Darwin' (AMST 283) introduced in Chap. 4. For more information, including an overview of the syllabus and accompanying radio and TV broadcasts see the Open University Digital Archive page at https://www.open.ac.uk/library/digital-archive/module/xcri:A381/ *(last accessed 19/02/2021)*.

Table 7.1 All BBC broadcasts featuring creationism (1925–2009)

Year	*Channel*	*Series/production*	*Episode*
1981	BBC 2	Horizon	Did Darwin get it wrong?
	BBC 2	Horizon	Did Darwin get it wrong? (REPEAT)
	Radio 4		In the Beginning ...
	Radio 4	Woman's Hour	Science Simplified
	BBC 2	Open University	The Tennessee Evolution Trial
	BBC 2	Open University	The Tennessee Evolution Trial (REPEAT)
	BBC 1	Everyman	Genesis Fights Back
1982	Radio 4	Open University	Darwin at Westminster
	BBC 2	The Man Alive Debate	Adam and the Ape
	BBC 1	Everyman	Genesis Fights Back (REPEAT)
	Radio 3		It Ain't Necessarily So
	BBC 2	Open University	The Tennessee Evolution Trial (REPEAT)
	BBC 2	Open University	The Tennessee Evolution Trial (REPEAT)
1983	BBC 2	Film	Inherit the Wind (REPEAT)
1984	BBC 2	Open University	The Tennessee Evolution Trial (REPEAT)
1986	BBC 2	Film	Inherit the Wind (REPEAT)
	Radio 3		The Blind Watchmaker
1987	BBC 2	Horizon	The Blind Watchmaker
1989	Radio 4	Soundings	In the Beginning
1992	Radio 4	The Great Monkey Trial	The State of Tennessee v John Thomas Scopes
1996	BBC 1	Everyman	Science Friction—Creation
1999	Radio 4	Leading Edge	*Report from Kansas*
2005	Radio 4	Crossing Continents	USA
2006	BBC 2	Horizon	A War on Science
2007	BBC 2	Film	Inherit the Wind (REPEAT)
2008	BBC 4	Film	Inherit the Wind (REPEAT)
2009	BBC 4	Film	Inherit the Wind (REPEAT)
	Radio 4	Leading Edge	Attitudes to Darwin
	Radio 4	The Saturday Play	The Great Tennessee Monkey Trial

List compiled using BBC Genome; where repeats aired in the same year, they have been omitted. List only includes those broadcasts dealing with creationism within an anti-evolutionary framework. For example, the list excludes those that consider creation within a purely religious context (e.g. "The Creation of the World", *In the Beginning*, January 18, 1970, BBC One), those that refer to non-evolutionary scientific theories of creation (e.g. *The Cosmic Creation*, June 1984, BBC Radio 3), and those that consider creation within an accommodationist evolutionary context (e.g. *Creation and the World of Science*, Brampton Lectures, written and presented by Arthur Peacocke, September 1978, BBC Radio 3)

historical event, from which, they discussed the subsequent development of the creationist and intelligent design movements. In a similar vein, the 1987 *Horizon* episode, "The Blind Watchmaker", began with evolutionary biologist and science populariser Richard Dawkins (b.1941) introducing William Paley's argument for intelligent design, the "watchmaker analogy". The documentary then jumped forward in time, with Dawkins appearing in the town of Glen Rose, Texas, where creationists have long claimed human footprints can be seen alongside fossilised dinosaur footprints preserved in the limestone of the nearby Paluxy River valley (Fig. 7.1). Much like in "Did Darwin Get It Wrong", the documentary confined creationism to the opening ten minutes, and Dawkins used pejorative and dismissive language throughout to refer to those who did not follow the orthodoxy of neo-Darwinian evolutionary thought.[45]

"The Blind Watchmaker" closely followed the content of Dawkins' 1986 popular book of the same name, and a BBC Radio 3 show about the book, again illustrating the close relationship between BBC science features and other popular science media. Further, the near-identical repetition of content across print—both the book itself and an extensive media campaign—and the radio and television shows highlights the importance of individual figures in popularising scientific narratives across different sectors of British society, via multiple popular mediums.[46]

As with "Did Darwin Get it Wrong?", "The Blind Watchmaker" only superficially covered the specifics of creationist ideology, using creationism as a narrative device to set up a dichotomous framing centred on conflict. Although made six years apart, both documentaries positioned themselves as responses to a recent rapid rise of scientific creationism in the US. Yet given the focus of their content, one cannot help but feel that they were more a defence of evolution and the hegemonic authority of science, rather than a direct deconstruction of creationist claims. Such unsymmetrical coverage of the two sides is in keeping with sociologist, Simon Locke's

[45] *Horizon*, "The Blind Watchmaker", written, presented and co-produced by Richard Dawkins, January 19, 1987, BBC 2. Those interested can watch the documentary online at https://www.dailymotion.com/video/x223a3n *(last accessed 19.06.21)*.

[46] By the 1980s this approach, which saw the alignment of science content across different media, most regularly alongside a popular science book was common. Over ten years earlier, the same approach was used alongside Dawkins' *The Selfish Gene* (1976), which also featured as the focus of a *Horizon* episode. *Horizon*, "The Selfish Gene", November 15, 1976, BBC Two; and Horizon: The Selfish Gene (Science Features), T63/109/1, BBC-WA.

Fig. 7.1 Richard Dawkins highlights a footprint, claimed by Creationists to be that of early man, by the Paluxy River near Glen Rose, Texas. Screenshot taken from Horizon "The Blind Watchmaker", BBC 2, 1987, 7:03.

argument that we cannot see the emergence of scientific creationism simply as a revival of creationism. However reactive the adoption of bombastic language and reductive narratives in science broadcasts was,[47] we must

[47] While "The Blind Watchmaker" documentary came relatively early in Richard Dawkins science communication career, the bombastic tone and reductionist view of science evident in this episode of *Horizon* would go on to become a hallmark of his style, and latterly of the wider "New Atheist" movement. For a later much more explicitly polemic example see: *The Root of All Evil?*, written by Richard Dawkins, produced by Alan Clements, Channel 4, January 2006: https://vimeo.com/27692770 *(last accessed 23/06/2021)*. For a review of a 1996 Dawkins' TV episode that highlighted these qualities see: Michael Walsh, "Television", *The Tablet*, October 5, 1996, 21–22; and for more on the language of the New Atheist movement see: Stephen Ledrew, 'Scientism, Humanism, and Religion: The New Atheism and the Rise of the Secular Movement' (York University, PhD Thesis, 2013), 56–96.

also acknowledge that this response amounted to "a revival of the crusade for evolution".[48]

In response to the 2005 Dover, Pennsylvania Intelligent Design Trial, which challenged the regional school board's inclusion of Intelligent Design on the curriculum, *Horizon* again visited the subject of creationism in 2006.[49] Again utilising weaponised and pejorative language, "A War on Science" gave viewers a more detailed account of American creationism than prior attempts. The documentary presented a narrative of continuous conflict between creationist thought and evolutionary theory, from nineteenth-century clashes, through the 1925 Scopes trial, via the 1987 Supreme Court ruling on creationism in the classroom, and up to the 2005 court case. This time, other recognisable figureheads including David Attenborough, joined Dawkins, "in an attempt to halt the movement's rise and keep religion out of science".[50] This fear-centred framing, 25 years after "Did Darwin Get it Wrong", was again justified as a response to the immediate threat posed by the growth of creationism, and in particular the Intelligent Design movement.

It is not particularly surprising that these three episodes of *Horizon*, the world's longest-running TV science series, were dismissive in their treatment of the pseudoscience of creationism. More illuminating, however, is how all three attempted to communicate the sometimes-complex evolutionary science that they featured. Despite utilising a range of creative visual tools and being open about the limitations of contemporary evolutionary science, all three programmes adopted what science communication scholars would recognise as a "deficit model" of science communication.[51] All three documentaries presumed that creationists, and those viewing at home who may sympathise with their views, only held such a worldview because they were ignorant of the real science behind evolutionary theory. Nowhere was this deficit model approach more apparent than in "The Blind Watchmaker", as Dawkins lectured the audience on the real scientific mechanisms behind macro-evolutionary

[48] Simon Locke, *Constructing the Beginning: Discourses of Creation Science* (Taylor & Francis, 1998), 63.

[49] *Horizon*, "A War on Science", produced by James van Der Pool, January 26, 2006, BBC 2.

[50] "Horizon", *Radio Times*, Issue 4268, January 19, 2006, 94.

[51] For an introduction to the deficit model and an overview of its widespread scholarly critique see Chap. 4 in Jane Gregory and Steven Miller, *Science In Public* (Basic Books, 2000).

processes, rebutting creationist's claims, and in the process seemingly believing he was shattering the assumptions of viewers who may have held creationist views. It is very unlikely that those who held a formalised anti-evolution position prior to the show, especially scriptural literalists like scientific creationists, would be watching *Horizon* in the first instance.[52] Rather, what these episodes of *Horizon* actually did was work to reinforce and demarcate the boundary of proper science. The documentaries appealed to those who already consumed science media, and for whom science was often a central component of their social identity.[53]

In contrast, programmes made directly by the BBC Religion Department gave a more balanced treatment of creationism, expecting of their audience a more nuanced religious literacy. In particular, the episodes of *Everyman* in 1981 and 1996 (Table 7.1) struck a pluralistic tone, featuring an accommodation of personal faith that did not simply dismiss fundamentalist groups outright. Unlike the *Horizon* episodes, the 1996 *Everyman* episode on creation, part of a trilogy called "Science Friction", attempted to unpack the religious, moral and ethical arguments bound up in opposition to evolution, managing successfully to place them in their geopolitical context. The episode again focused on the Intelligent Design movement, but this time used the 1995 Anti-Evolution Bill in Nashville, Tennessee, as its topical news hook. The episode was the only one in Table 7.1 to go beyond stereotypes. It presented people with views across a spectrum of possible beliefs, from mixed views among a group of retirees in a café, in Dayton, Tennessee, to a petition organised by local schoolchildren who feared restricted future job prospects if they did not understand

[52] While reliable data on the prevalence of literal creationist views in the UK during the 1980s is limited, a more recent 2015 survey showed that only around 9% of the British public have a literalist creationist worldview. That is, that when surveyed they agreed with the statement, "Humans and other living things were created by God and have always existed in their current form". Fern Elsdon-Baker et al., "Results of Major New Survey on Evolution," press release, Science and Religion: Exploring the Spectrum (blog), September 5, 2017, https://sciencereligionspectrum.org/in-the-news/press-release-results-of-major-new-survey-on-evolution/ *(last accessed 21/06/21)*. For more on some of the challenges assessing and understanding public attitudes towards evolution in the UK see: Fern Elsdon-Baker, "Creating hardline 'secular' evolutionists: The influence of question design on our understanding of public perceptions of clash narratives between evolutionary science and belief" in Fern Elsdon-Baker and Bernard Lightman (Eds), *Identity in a Secular Age: Science, Religion, and Public Perception* (University of Pittsburgh Press, 2020).

[53] For more on science as a part of social identity and these themes in a range of contexts see Elsdon-Baker and Lightman.

both creationist and evolutionary positions.[54] Yet, while some of the "everymen" of the show's title presented a range of conflicting, confused and ultimately real beliefs, the experts featured did not veer from the expected normative position on their respective side of the creationism and evolutionary science divide. As noted in one religious newspaper, the show did not even give a platform to any Christians who believed in evolution, reinforcing the stereotype that evolution was necessarily atheistic and thus inadvertently giving credence to the creationists' own fears.[55]

Reflecting its origin within the Religion Department and its billing on the more highbrow station BBC Radio 4, *In the Beginning* (1986) included the most diverse views on the debate between creationism and evolution. Including theologians, ministers, scientists and a historian, the show was the only one broadcast in the 1980s (Table 7.1) that featured a non-Christian perspective on the relationship between religious creation stories and contemporary understandings of science.[56] *In the Beginning* was also the only programme about evolution and creationism on the BBC to feature representatives of the British creationist movement, featuring Minister in the Church of Scotland and Chairman of the Biblical Creation Society, Nigel Cameron, their President, material scientist Edgar Andrews, and one of their members, biochemist David Gower. Gower, then a Reader in Biochemistry at Guy's Hospital in London, rather alarmingly outlined how he had been visiting universities in the UK to speak to students about his anti-evolutionary religious beliefs.[57]

[54] *Everyman*, "Science Friction: Creation", produced by Paul Sapin, September 8, 1996, BBC 1. The subsequent two episodes in the trilogy were on "Genes" and "Miracles". The title, "Science Friction" was first used by the BBC for a Radio 4 series, which ran from 1990 to 1993 and looked at the appliance of science in daily life. Those interested can watch "Science Friction – Creation" online at: https://www.youtube.com/watch?v=Ul8cRbIv6eE *(last accessed 21/06/21)*.

[55] Michael Walsh, "Television", *The Tablet*, September 14, 1996, 24.

[56] Using Genesis to connect the Bible and the Torah, the broadcast interviewed Rabbi Malcolm Wiseman who explained that while for some fundamentalists evolutionary science may present issues, the majority of Jewish people interpreted Genesis as an allegorical story, and as such, evolutionary science did not cause any issues. Like the OU broadcasts on this subject in the period, the historian featured was James Moore. *In the Beginning*, presented by Robert Foxcroft, produced by Sarah Widdows, July 14, 1981, BBC Radio 4. Transcript on microfilm at BBC-WA. All BBC copyright content reproduced courtesy of the British Broadcasting Corporation. All rights reserved.

[57] Gower went on to become a Professor of Biochemistry at the University of London and throughout his career remained a vocal anti-evolutionist and creationist. For more on his own personal views on the subject see Chapter 28 in: John F. Ashton, *In Six Days: Why Fifty 50 Scientists Choose to Believe in Creation* (New Leaf Publishing Group, 2001). For more on

All of the programmes introduced thus far, despite their outright rejection of the truth claims of creationism, gave a platform to a predominantly US phenomenon. The BBC, as a media producer that reacts to global events and trends, was covering the latest controversy coming out of the US. However, media producers cannot control the effect of their coverage, and thus inadvertently they may amplify a narrative they merely set out to report. Broadcast on a weekday evening on BBC Radio 4, *In the Beginning* gave a, albeit brief, platform to the UK-based Biblical Creation Society, which at the time only had around 700 members nationally.[58] Although the other guests all discredited scientific creationism, the inclusion of British proponents of scientific creationism—an extremely fringe position among British Christians—may have influenced those listening at home who were already sympathetic to this worldview.[59] Thus in attempting to report a US-led phenomenon in a responsible, fair and balanced way as per the BBC's public charter, the organisation gave a platform to an extreme minority view and amplified a contemporary narrative of conflict between evolution and religion that was not representative of mainstream trends in the UK.[60]

Nigel Cameron, the Biblical Creationist Society, and Edgar Andrews see: Numbers, *The Creationists*, 357–360.

[58] By the mid-1980s, the Biblical Creation Society had 750 members. Numbers, *The Creationists*, 358.

[59] The 2015 SRES/YouGov survey found that 16% of those who identified as religious or spiritual held the creationist belief that, "Humans and other living things were created by God and have always existed in their current form". Remembering that this top-level data includes those from all religious groups and non-traditional spiritual beliefs, and only an extreme fraction of this 16% is likely to be part of formalized anti-evolution organizations. Fern Elsdon-Baker et al., "Results of Major New Survey on Evolution," press release, Science and Religion: Exploring the Spectrum (blog), September 5, 2017, https://sciencereligion-spectrum.org/in-the-news/press-release-results-of-major-new-survey-on-evolution/ *(last accessed 21/06/21).*

[60] Although from the 1980s onwards the BBC gave a small platform to creationist ideas, the role of the British media in promoting the ideology was still extremely limited compared to the central role broadcast media played in debates in the US. For a clear juxtaposition of these national contexts, see two films aired almost back to back on the US Cable-Satellite Public Affairs Network in 2014. The first was a wide-ranging talk on the "Media Coverage of Religion", given by the BBC Religious Affairs correspondent, Jane Little covering her experiences during the 1990s and beyond, which does not mention creationism; and the second is the infamous "Evolution Versus Creationism Debate", featuring science broadcaster Bill Nye and Christian fundamentalist and founder of *Answers in Genesis*, Ken Ham. Both broadcasts, in their originally aired format and scheduling, can be viewed online at

Religious Metaphors on Science Broadcasts

Across the second half of the twentieth century, BBC science content, particularly content on evolution, increasingly adopted religious language and metaphors, not just during broadcasts as seen in Chap. 6, but also in the titles and advertised listings used to attract an audience. By contrast, in the first decades of the BBC's history, biblical phrases such as. "serpent in the Garden of Eden" and "In the beginning …" were reserved exclusively for religious programming looking at human origins through the lens of Genesis. However, by the 1960s, broadcasters increasingly deployed this religious language metaphorically on science shows that dealt with the origins of life and humans.[61] Reflecting the renewed interest in discussions around religious literalism and evolution, this trend saw a marked uptick during the 1980s. Episodes in major science series, such as palaeoanthropologist Richard Leaky's *The Making of Mankind* (1981) and *Imagined Worlds* (1982) prominently used biblical language.[62] In the latter, an episode that dealt with how we might "escape the tyrannies of inheritance" was provocatively titled "Serpent in the Garden of Eden".[63] Such language did not just feature on science documentaries, but also in dramatised pieces, such as the experimental series *The Scientists* (1980).[64] The categorical co-opting of sacred language, previously preserved for religious use, across a range of scientific programmes reflects secularisation trends in Britain during the period, and alerts us to the role that science as a growing cultural authority played in these changes. However, we must be careful when drawing links between such small semantic shifts and such

https://archive.org/details/CSPAN_20140223_044300_Evolution_Versus_Creationism_Debate *(last accessed 20/06/2021).*

[61] For an early example, see the first episode of a ten-part television series on molecular biology: *What is Life?*, "In the Beginning was the Cell", presented by Professor Asher Korner and Dr Geoffrey Eglinton, directed by Mary Hoskins, October 9, 1967, *BBC One.*

[62] *The Making of Mankind*, "In the Beginning", written and presented by Richard Leakey, February 3, 1982, BBC Two; and *Imagined Worlds*, "Serpent in the Garden of Eden", March 8, 1982, BBC Two.

[63] "Imagined Worlds", *The Radio Times*, Issue 3043, March 4, 1982, 43.

[64] This three-part series of dramatised debates, originally aired for schools, featured a computer hologram playing scientific intellectuals from history brought to life to challenge the beliefs of modern scientists. In the second episode, "In the Beginning", the computer called up Huxley and Wilberforce in an attempt to reconcile their differences on scientific responsibility. *The Scientists*, "In the Beginning", written by Jim Hawkins, directed by Michael Coyle, May 13, 1980, BBC Two; and Elizabeth Cowley, "The Scientists", *The Daily Mail*, May 6, 1980, 23.

macroscale trends. As revisionist accounts of British secularisation remind us, large-scale societal change is rarely uniform and as neatly bound as one hegemonic authority replacing another.[65] Indeed, to return to this small example, as well as being adopted by science broadcasts, religious phrases referring to human origins continued to be employed prominently in religious broadcasts throughout the 1980s.[66] As with *Horizon's* repeated inclusion of scientific creationism purely as a novel narrative hook, in instances when science shows used religious metaphors, they rarely engaged with the scriptural background of the phrases. By the 1980s, British science broadcasters were comfortable using religious language in a secularised context, and did not see religious contestation of scientific ideas as a threat to either their position as science communicators, scientific knowledge itself or science's authoritative position in modern British society.

Conclusion

Following one of the more recent re-emergences of the latest iteration of the waterside ape theory, in 2013 a senior editor from *Nature*, the palaeontologist Henry Gee, argued that we should consider the aquatic ape hypothesis as a form of creationism.[67] While Gee was convincing, if a little overzealous, in his succinct articulation of the gaps and holes in the theory, he was less clear on why such an absence of evidence meant the theory was equivalent to an anti-evolutionist position based on intervention by a supernatural God.

In this chapter, we have seen that while the BBC had a relatively long history of broadcasts that dealt with the philosophical, ethical and societal implications of evolutionary science, up until the 1980s, content rarely contained anti-evolutionist positions. In fact, the content that most directly criticised elements of evolutionary science came from science broadcasts themselves. On these very limited occasions, scientific and

[65] Callum G. Brown, *The Death of Christian Britain* (Routledge, 2013).

[66] See, for example, the 1984 BBC airing of the religious epic film *The Bible ... In the Beginning* (1966). *The Bible ... In the Beginning*, screenplay by Christopher Fry, directed by John Huston, April 12, 1984, BBC Two. The film is available online: https://archive.org/details/The_Bible_In_The_Beginning *(last accessed 23/02/21)*.

[67] Henry Gee, "Aquatic apes are the stuff of creationism, not evolution", *The Guardian*, May 7, 2013, https://www.theguardian.com/science/occams-corner/2013/may/07/aquatic-apes-creationism-evolution *(last accessed 17/02/21)*.

pseudo-scientific theories were given airtime to critique the neo-Darwinian consensus of the day. We could view these instances as the BBC errantly giving a platform to pseudoscience, as did the scientists who criticised Attenborough's broadcasts on the aquatic ape theory in 2016. However, given that the broadcasts on the aquatic ape theory were based on peer-reviewed science, we can also understand these examples as opportunities for exposing a wider audience to the processes of science, and scientific views that diverge from the consensus position.

The coverage afforded to creationist ideologies in the 1980s, while initially triggered by journalistic interest in the growth of scientific creationism in the US, was perpetuated by the BBC's policy on impartiality and a desire to show opposing points of view. The framing used was one centred on conflict, and often utilised fear, particularly in science broadcasts on the subject that were purportedly worried such fundamentalist ideas could grow significantly in the UK. Inadvertently, such coverage acted as a potential conduit for these ideas, briefly giving a platform to extremely marginal religious groups and their anti-evolutionist ideas. While this may have appealed to a very small portion of the British population who already sympathised with creationist ideas, the fact that these programmes very rarely dealt in depth with creationist ideology alerts us to the fact that this was really a defence of science. Science broadcasts like *Horizon* used creationism both as an audience hook, and to highlight the superior explanatory power of evolutionary science. Broadcasters' defence of science was not based on a real fear that religion might outcompete scientific knowledge, but rather on ensuring there was a clear demarcation between evolutionary science and fundamentalist creationist groups, who were increasingly adopting scientific language.

When taken as a whole, the BBC broadcasts that included content on scientific creationism did not just attempt to impart scientific knowledge, or even teach about the process of science, but rather presented evolutionary science as a belief system that was superior to literalist religious traditions. As with the general mode of speaking about evolutionary science that we have seen develop across the prior chapters of this book, this was a worldview that was not explicitly atheistic. There was space for religious belief as long as it was malleable, and where necessary, conceded to scientific hegemony. By the 1980s, this approach, so directly shaped by earlier scientific humanist ideology, was embedded in the very formats of popular science broadcasting in the UK. Science, with an evolutionary framework as a central tenet, was increasingly becoming a key part of cultural identity in the UK. Not a well-defined scientism, as perhaps Huxley and peers in the 1930s had imagined, but a loser identity, in which individuals connected popular broadcasts on subjects like evolution, with science fiction, and a secular humanist worldview.[68]

References

Ashton, John F. *In Six Days: Why Fifty 50 Scientists Choose to Believe in Creation*. New Leaf Publishing Group, 2001.

Boykoff, Maxwell T. *Who Speaks for the Climate?: Making Sense of Media Reporting on Climate Change*. Cambridge University Press, 2011.

Brown, Callum G. *The Death of Christian Britain*. Routledge, 2013.

Calder, Nigel. *The Life Game: Evolution and the New Biology*. Viking Press, 1974.

Caulfield, Timothy. 'Popular Media, Biotechnology, and the Cycle of Hype The Mass Media's Influence on Health Law and Policy Symposium'. *Houston Journal of Health Law & Policy* 5, no. 2 (2005): 213–234.

Cowley, Elizabeth. "The Scientists", *The Daily Mail*, May 6, 1980, 23.

———. "Pick of the Day: Horizon." *The Daily Mail*, March 30, 1981.

Elsdon-Baker, Fern, and Bernard Lightman. *Identity in a Secular Age: Science, Religion, and Public Perception*. University of Pittsburgh Press, 2020.

Fern Elsdon-Baker et al., "Results of Major New Survey on Evolution," press release, Science and Religion: Exploring the Spectrum (blog), September 5, 2017, https://sciencereligionspectrum.org/in-the-news/press-release-results-of-major-new-survey-on-evolution/.

Foley, Robert, and Marta Mirazón Lahr. 'The Role of "the Aquatic" in Human Evolution: Constraining the Aquatic Ape Hypothesis'. *Evolutionary Anthropology: Issues, News, and Reviews* 23, no. 2 (2014): 56–59. https://doi.org/10.1002/evan.21405.

Gee, Henry. "Aquatic apes are the stuff of creationism, not evolution", *The Guardian*, May 7, 2013, https://www.theguardian.com/science/occams-corner/2013/may/07/aquatic-apes-creationism-evolution.

Gregory, Jane, and Steven Miller. *Science In Public*. Basic Books, 2000.

Hardy, Alister. *The Living Stream*. Harper & Row, 1965.

———. "Was Man More Aquatic in the Past?, *The New Scientist*, March 17, 1960, 642–645.

Harman, Oren, and Michael R. Dietrich. *Outsider Scientists: Routes to Innovation in Biology*. University of Chicago Press, 2013.

"Horizon", *Radio Times*, Issue 4268, January 19, 2006, 94.

"Inherit the Controversy", *Newsday* (New York), March 17, 1996, 85.

[68] See Elsdon-Baker and Lightman, *Identity in a Secular Age*.

Korn, Eric. "Remainders", *The Times Literary Supplement*, May 15, 1981, 541.
Leach, Edmund. *A Runaway World? The B.B.C. Reith Lectures 1967*. Oxford University Press, 1968.
Ledrew, Stephen. 'Scientism, Humanism, and Religion: The New Atheism and the Rise of the Secular Movement' (York University, PhD Thesis, 2013).
"Letters", *New Scientist*, January 25, 1973, 210–211.
Lewis, John, and Bernard Towers. *Naked Ape or Homo Sapiens?* Garnstone Press, 1972. http://archive.org/details/nakedapeorhomosa00lewi.
Lightman, Bernard. 'Darwin and the Popularization of Evolution'. *Notes and Records of the Royal Society*, 20 March 2010. https://doi.org/10.1098/rsnr.2009.0007.
Locke, Simon. *Constructing the Beginning: Discourses of Creation Science*. Taylor & Francis, 1998.
Macbeth, Norman. *Darwin Retried: An Appeal to Reason*. Gambit Incorporated, 1971.
Maynard Smith, John. "Counsel for the defence," *New Scientist*, 22 August 1974, 476.
Milam, Erika Lorraine. *Creatures of Cain: The Hunt for Human Nature in Cold War America*. Princeton University Press, 2019.
Moore, Randy. 'Creationism in the United States: VII. The Lingering Impact of "Inherit the Wind"'. *The American Biology Teacher* 61, no. 4 (1999): 246–250. https://doi.org/10.2307/4450666.
Morris, Colin. *God-in-a-box: Christian strategy in the television age*. Hodder and Stoughton, 1984.
Morris, Desmond. *The Naked Ape: A zoologist's study of the human animal*. Jonathan Cape, 1967.
Numbers, Ronald L. *The Creationists: From Scientific Creationism to Intelligent Design*. Expanded ed edition. Harvard University Press, 2006.
"Personal Choice." *The Times*, March 30, 1981.
Roberts, Alice, and Mark Maslin. "Sorry David Attenborough, we didn't evolve from 'aquatic apes' – Here's why", *The Conversation*, September 16, 2016, https://theconversation.com/sorry-david-attenborough-we-didnt-evolve-from-aquatic-apes-heres-why-65570.
Rutherford, Adam. *Humanimal: How Homo Sapiens Became Nature's Most Paradoxical Creature—A New Evolutionary History*. The Experiment, 2019.
Shapiro, Adam R. *Trying Biology: The Scopes Trial, Textbooks, and the Antievolution Movement in American Schools*. University of Chicago Press, 2013.
Straalen, Nico M. van. 'The Naked Ape as an Evolutionary Model, 50 Years Later'. *Animal Biology* 68, no. 3 (1 January 2018): 227–246. https://doi.org/10.1163/15707563-17000167.
Tambiah, Stanley Jeyaraja, and Stanley J. Tambiah. *Edmund Leach: An Anthropological Life*. Cambridge University Press, 2002.

"Today's TV and sound," *The Observer*, October 10, 1965, 22.

Tuttle, Russell H. 'Review of The Naked Ape: A Zoologist's Study of the Human Animal; The Apes: The Gorilla, Chimpanzee, Orangutan and Gibbon – Their History and Their World, by Desmond Morris and Vernon Reynolds'. *American Anthropologist* 70:6 (1968): 1238–1240.

Ungureanu, James C. *Science, Religion, and the Protestant Tradition: Retracing the Origins of Conflict*. University of Pittsburgh Press, 2019.

Walsh, Michael. "Television", *The Tablet*, September 14, 1996a, 24.

———. "Television", *The Tablet*, October 5, 1996b, 21–22.

Ward, Bob. "The BBC is sacrificing objectivity for impartiality in its coverage of climate change", *British Politics and Policy at LSE*, July 11, 2012. http://eprints.lse.ac.uk/id/eprint/48416.

Woodruff, Douglas. "Talking at Random", *The Tablet*, December 2, 1967, 23.

CHAPTER 8

Remembering or Deifying? The Darwin Anniversaries of 1959 and 2009

[T]he very notion of "Darwinism" is misleading, suggesting not a programme for research but a set of dogmas which bold spirits will want to rebel against. Certainly Darwin constructed an edifice: but that edifice is a laboratory, not a temple.
—Eric Korn, review of "Did Darwin Get it Wrong?" 1981 (Eric Korn, "remainders", *The Times Literary Supplement*, May 15, 1981, 541)

In the winter of 1959, biologist and regular on BBC broadcasts, Peter Medawar delivered the annual *Reith Lectures* on "The Future of Man".[1] Over fifteen years after the wartime series, *Reshaping Man's Heritage* (1943) and eight years before Edmund Leach's "A Runaway World?" *Reith Lectures* (1967), on this occasion it was Medawar's turn to speak about the implications of biological knowledge for humanity's future.[2]

[1] *The Reith Lectures*, "The Future of Man", written and presented by Peter Medawar, November to December 1959, BBC Home Service. The lectures were repeated the following day on the BBC Third Programme, serialised in the BBC magazine *The Listener*, and published as a book the following year. *The Future of Man: The BBC Reith Lectures 1959* (Basic Books Inc., 1960).

[2] The six titles of Medawar's lectures: The Fallibility of Prediction, The Meaning of Fitness, The Limits of Improvement, The Genetic System of Man, Intelligence and Fertility and The Future of Man.

A. Hall, *Evolution on British Television and Radio*, Palgrave Studies in Science and Popular Culture,
https://doi.org/10.1007/978-3-030-83043-4_8

The six talks covered areas chosen to answer the question: can humans go on evolving as they have done in the past? Medawar repeatedly drew a distinction between proper socio-biology and misapplications of such scientific knowledge—"bad judgements based upon bad biology"—in an attempt to present "a 'humane' solution to the problem of eugenics".[3] Medawar dealt with this most comprehensively in the final lecture. In language perhaps a little complex for the general listener, he outlined the differences between endo-somatic hereditary—the genetic processes we share with other species—and the exo-somatic hereditary processes unique to humanity, such as the intergenerational transmission of knowledge via the written word.[4] Much more measured in tone than both Huxley and Leach's broadcasting style, the lectures nonetheless attempted to situate humanity's societal and cultural evolution within the purview of biological study.

The BBC chose Medawar for the *Reith Lectures* in 1959 to coincide with the centenary of the publication of *On the Origin of Species* and the 150th anniversary of Darwin's birth. As well as a plethora of conferences and celebrations globally, the anniversary saw the BBC produce a wide range of broadcasts across both radio and television. This chapter first introduces the general celebrations in 1958–1959, and then outlines the broadcast content created for the anniversary, highlighting how scientists used the commemorations as an opportunity to promote a new unified vision of biology. The chapter then explores how broadcasters have used subsequent Darwin anniversaries as a hook for broadcast content, and over the decades following 1959, how such content contributed to the emergence of a mythic narrative of Darwin's life and influence. The chapter then turns to the huge number of broadcasts made under the auspices of the 2009 Darwin Anniversary, and uses them as a vantage point on which to reflect on the major themes that have emerged over the course of the preceding chapters.

[3] Peter Medawar, *The Future of Man: The BBC Reith Lectures 1959* (Basic Books Inc., 1960), 99.

[4] The final lecture in the series can be listened to online at https://www.bbc.co.uk/sounds/play/p00hg13w *(last accessed 01/06/21).*

The Darwin Sesquicentennial Celebrations 1958–1959

In 1909, 100 years since Darwin was born and 50 years after the publication of *On the Origin of Species*, celebratory events took place in New York, New Zealand and most notably Cambridge, where over 400 scientists and dignitaries from 167 different countries gathered.[5] In 1909 Darwin's scientific legacy was still unclear, particularly as his theory of natural selection was not yet widely accepted as the main driving mechanism of evolution. As well as the familiar academic activities, such as speeches, lectures and the bestowing of honorary degrees, the Cambridge event also included an exhibition of Darwin memorabilia and his old rooms at Christ's College were opened to visitors.[6] The 1909 celebrations have been analysed by historians of science as the culmination of an early process of appropriating Darwin as a hero of science, which grew out of Darwin's public image in his own lifetime. Bolstered by a slew of early hagiographic biographies, Darwin's legacy became increasingly about his painstaking and rigorous powers of observation, his trustworthiness and position as a moral exemplar, and his personal scientific and religious beliefs.[7]

By 1959, 150 years after Darwin's birth and 100 years since *Origin*'s publication, a lot had changed in the biological sciences. At the turn of the century, Gregor Mendel's work on hereditability had been rediscovered and natural selection was now a widely accepted mechanism of evolution. Through the work of the English statistician and biologist R. A. Fisher (1890–1962), among others, Mendelian genetics was combined with Darwinian natural selection to unite the then disparate branches of biology, ultimately bringing together subjects as diverse as molecular biology, ecology and palaeontology. Julian Huxley played an integral role in synthesising and popularising these new combined approaches in biological

[5] This section was adapted with permission from: Alexander Hall, Tom Kaden, and Fern Elsdon-Baker, "Darwin Day: Celebrating Without Deifying", *Science and Religion: Exploring the Spectrum Blog*, February 2016.

https://sciencereligionspectrum.org/in-the-news/darwin-day-celebrating-without-deifying/ *(last accessed 02/06/2021).*

[6] John van Wyhe, "1909: The first Darwin centenary", *The Complete Work of Charles Darwin Online*, 2002, http://darwin-online.org.uk/1909.html *(last accessed 02/03/21).*

[7] Bernard Lightman, 'The Many Lives of Charles Darwin: Early Biographies and the Definitive Evolutionist', *Notes and Records of the Royal Society* 64:4 (2010): 339–358, https://doi.org/10.1098/rsnr.2009.0059.

evolution, coining the term "the modern synthesis" to encapsulate developments in 1942.[8]

The 1959 anniversary saw international activities on an unprecedented scale, with events across the world, from Australia to Brazil to the Soviet Union. In a break from prior Darwin anniversaries, the major commemorative event did not take place in the UK. In November 1959, over five days at the University of Chicago, 2500 registered participants met for a full academic conference. There was a full public programme of talks and events, and the conference attracted a huge amount of media attention. The celebration did not focus particularly on Darwin's work or ideas, and was instead largely about promoting the new unified field of evolutionary biology. The location and focus helped to ensure that the US was seen as central in the field, and that subjects from beyond the traditional boundaries of scientific endeavour, such as anthropology, were brought into the evolutionary biological project.[9] The centenary of Darwin's major publication presented an opportunity to frame recent developments and celebrate the unifying of biology and evolution. As the historian of biology Vassiliki Betty Smocovitis has shown, celebrants at the 1959 Chicago events "rejoiced and at the same time reified their own new identities as sons—and, literally, as grandsons of the protagonists of the original story".[10]

One of the most rousing speeches delivered during the 1959 celebrations was Julian Huxley's convocation address, "The Evolutionary Vision".[11] Covering territory familiar to anyone who had encountered his popular broadcasts or books on the subject over the prior decades, the speech saw Huxley step-up the rhetoric used to promote his scientific humanist worldview. In his expression of evolution as a universal phenomenon, which in addition to the natural sciences should also encompass the

[8] Julian Huxley, *Evolution, the Modern Synthesis* (Allen & Unwin, 1942); and Vassiliki Betty Smocovitis, *Unifying Biology: The Evolutionary Synthesis and Evolutionary Biology* (Princeton University Press, 1996), 97–153.

[9] Vassiliki Betty Smocovitis, 'The 1959 Darwin Centennial Celebration in America', *Osiris* 14 (1999): 274–323.

[10] Smocovitis, 'The 1959 Darwin Centennial Celebration in America', 321.

[11] The full proceedings of the centennial celebrations can be read online, Huxley's speech is in volume 3, pages 249–261. Sol Tax and Charles Callendar, *Evolution after Darwin: the University of Chicago Centennial*, Volume III (University of Chicago Press, 1960), https://archive.org/details/evolutionafterda03taxs/page/n5/mode/2up *(last accessed 02/06/21)*.

human mind and culture, Huxley was more forthright than the tempered exposition of Peter Medawar's 1959 *Reith Lectures*. In both Huxley's speech and Medawar's broadcasts there is a common articulation of a distinction between what Huxley called biological and psychosocial evolution—Medawar's endo-somatic and exo-somatic hereditary processes.[12] Like Medawar, who situated the exo-somatic processes of cultural transmission within macro-biological processes, Huxley also saw the institutions of humanity's social-political world as important drivers of *Homo sapiens'* evolutionary adaptation. However, unlike Medawar, Huxley was comfortable extending his language beyond the safe confines of biology, using secular humanism to draw a link between the species' inner biology and external social structures. Huxley believed that evolutionary humanism was the vehicle that could take humanity's ethics and worldview forward, forming a new emergent religion:

> The emergent religion of the near future [...] should be able, with our increased knowledge of mind, to define our sense of right and wrong more clearly so as to provide a better moral support; it should be able to focus the feeling of sacredness onto fitter objects, instead of worshiping supernatural rulers, so as to provide truer spiritual support, to sanctify the higher manifestations of human nature in art and love, in intellectual comprehension and aspiring adoration, and to emphasize the fuller realization of life's possibilities as a sacred trust.[13]

Unsurprisingly, given his attack on traditional religious institutions, when the American popular press picked up Huxley's speech, it caused anger; with the British Embassy receiving one letter that called for Huxley's immediate deportation.[14]

[12] Krishna R. Dronamraju and Joseph Needham, *If I Am to Be Remembered: Correspondence of Julian Huxley* (World Scientific, 1993), 132–133.

[13] Julian Huxley, "The Evolutionary Vision: Convocation Address", in Sol Tax and Charles Callendar, *Evolution after Darwin: the University of Chicago Centennial*, Volume III (University of Chicago Press, 1960), 260.

[14] Smocovitis, 'The 1959 Darwin Centennial Celebration in America', 314. Huxley picked up this argument in his opening chapter in the 1961-collected edition, *The Humanist Frame* (Harper and Brothers, 1961). As well as featuring the writing of prominent scientists, including Jacob Bronowski, the book contained chapters by many prominent humanist thinkers of the period, including a closing chapter, "Human Potentialities" by Julian's brother, the author Aldous Huxley.

Although anti-evolution sentiment particularly that based on fundamentalist religious beliefs was much less prominent in British society (Chap. 7), Huxley would not have been allowed to deliver such an explicitly anti-Christian message on the BBC. While by 1959, reflecting wider trends of secularisation in British society, the BBC had begun to broaden the scope of the religious and non-religious worldviews broadcast; any content that directly questioned the authority of traditional Christianity was still widely censored.[15] By comparison, Medawar's *Reith Lectures* avoided any direct reference to the future role religion may play in society, and stuck to scientific language and subject matter throughout.[16] However, both Medawar's lectures and Huxley's Chicago speech were similarly aligned in their philosophical grounding, with both ultimately advocating an evolutionary-humanist future for humanity. Again highlighting, as we saw in Chap. 2, how prominent public intellectuals of the period were able to use science broadcasting to bypass religious censorship to air humanist, atheist and other secular worldviews.

In the UK, the formal celebrations focused on the 1958 centenary of the joint reading of Darwin and Wallace's papers before the Linnaean Society. The main commemorations came under the auspices of the proceedings of the 15th International Congress of Zoology held in London from July 16 to 23, 1958. Although featuring many luminaries from the world of evolutionary biology and members of Darwin's family, Julian Huxley dominated proceedings.[17] The scale and prominence of commemorations held in the UK were somewhat hampered by Huxley's desire to be at the centre of planning and delivery, and were therefore unsurprisingly later overshadowed by the size, comprehensive programme

[15] See 'Central Religious Advisory Committee: Minutes File 1, 1958–1961', R6/21/8, BBC-WA, which includes deliberation and discussion of whether to broadcast content on humanism, a Jewish service once a year and reflections on the marked decline in religious faith among the British population. It is important to note that in this period the Central Religious Advisory Committee also oversaw religious content on the newly licensed ITV network.

[16] In 1949, the BBC turned down Julian Huxley for the *Reith Lectures*, when via the producer Mary Somerville he pitched a series of lectures on "Man and Evolution". Following extensive revisions, including a review by Henry Dale, the talks eventually aired as *The Process of Evolution*, written and delivered by Julian Huxley, October to November 1951, BBC Third Programme. Talks. 'Sir Julian Huxley, File II & III, 1950–1962', Rcont1, BBC-WA.

[17] H. R. Hewer and Norman Riley, *XVth International Congress of Zoology, London, 16–23 July 1958: Proceedings*, Volume 1 (XVth International Congress of Zoology, 1959); "Notes and News." *The Auk*, 75:3 (1958): 378–380.

and accompanying media attention of the Chicago conference.[18] Despite Huxley's attempts to control all major commemorative events in the UK, many academic societies and civic organisations still commemorated both the 1958 and 1959 anniversaries.[19] The BBC joined newspapers and publishers in capitalising on the ongoing popularity of both Darwin as a historical figure and evolution as a scientific subject, with a sequence of dedicated broadcasts during 1958–1959.[20]

The Darwin Centenary on Radio and Television

Prompted by Archibald Clow, senior producers at the BBC began discussing the Darwin Centenary in early 1956. What started as a proposal for a series of radio lectures was soon added to by the newly established Natural History Unit in Bristol, which proposed a Darwin special of *The Naturalist*, with plans growing in scale during 1956.[21] Without the dedicated coordination and oversight of science programming still being called for by external scientific societies, however,[22] each department pitched its own ideas piecemeal, causing the Controller of Programme Planning to warn, that "[t]his looks like the beginning of a snowball. Darwin centenary [is] obviously important—but is this really the way to tackle [it]?"[23] As with the wider celebrations in the UK, some involved felt that the looming presence of Huxley was hindering the diversity of programming the BBC was able to offer. As early as June 1956, a member of the Talks Department

[18] Smocovitis, 'The 1959 Darwin Centennial Celebration in America', 281–282.

[19] Among other activities, including involvement with the 1958 London conference, the Royal Society produced a special issue to coincide with *Origins*' centenary, and the Linnaean Society unveiled a plaque commemorating the reading of Wallace and Darwin's papers. Royal Society of London, *Notes and Records*, 14:1, 1959; and "Darwin-Wallace Plaque", *The Times*, July 5, 1958, 8.

[20] For examples of the wide-ranging print coverage the anniversaries generated see: "The World of Science: David (Cannon) and Goliath (Darwin)", *The Illustrated London News*, July 5, 1958, 32; Gertrude Himmelfarb, *Darwin and the Darwinian Revolution* (Chatto and Windus, 1959); and M. L. Johnson, Michael Abercrombie and G. E. Fogg (Eds), *New Biology 28*, January 1959 (Penguin).

[21] Desmond Hawkins (Head of West Regional Programming) to Archibald Clow, BBC Memo, November 23, 1956, R34/981, BBC-WA.

[22] Calls for more oversight and coordination of science broadcasting continued to be led by the Royal Society and the British Association for the Advancement of Science. 'Advisory Committees: Science Consultative Group, 1958–63', R6/239/1-BBC-WA.

[23] H. Rooney Pelletier (Controller, Programme Planning (Sound)) to Assistant Director Sound Broadcasting, BBC Memo, May 21, 1957, R34/981, BBC-WA. All BBC copyright content reproduced courtesy of the British Broadcasting Corporation. All rights reserved.

complained, that "[m]ost of the philosophers either are not interested in, or are frightened of, the subject: and most of the evolutionary biologists are either philosophically naïve, or are Huxleian".[24] While this comment was made in the specific context of trying to plan an anniversary broadcast on the philosophy of evolution, it hints at a wider understanding by BBC producers that Huxley's focus on celebrating the modern evolutionary synthesis came with his progressive humanist framing.

What emerged from these early deliberations was an extensive, but somewhat fragmented series of broadcasts spread over the different BBC radio and television channels throughout 1958–1959 (Table 8.1). At the heart of radio content for the Darwin Centenary was a series of lectures recorded with world-leading experts when they were in London for the International Congress of Zoology celebrations in July 1958. Although covering evolution from a number of disciplinary angles, reflecting the make-up of the conference, the majority of these shows featured evolutionary biologists directly involved with the modern synthesis.[25] For example, inspired by his contributions to a London Zoological Society pamphlet produced for the conference, embryologist and Director of the British Museum, Gavin de Beer provided the main history of science content in the radio celebrations.[26] Even broadcasts focused more on the philosophical, social and religious implications of Darwin's work featured evolutionary biologists. In fact, *Religion and Evolution* (1958), an attempt to explore "the 19th century controversy in religious matters precipitated by the theory of Natural Selection" *only* featured biologists, albeit of varying religious positions.[27] Hosted by Peter Medawar, this radio discussion and an episode of Christian Outlook a week later (Table 8.1) were both

[24] Barry Jackson (Talks Department) to Chief Assistant (Third Programme), BBC Memo, June 8, 1953, R34/981, BBC-WA. All BBC copyright content reproduced courtesy of the British Broadcasting Corporation. All rights reserved.

[25] In addition to those discussed here this series of radio broadcasts also featured among others, evolutionary biologist, Ernst Mayr ("The Real Origin of Species", May 20, 1959); geneticist and eugenicist C. D. Darlington ("The Control of Evolution in Man", January 9, 1959); and geneticist and eugenicist Ronald Fisher (*Evolution in Action*, "The Discontinuous Inheritance", July 8, 1958). See John Simons, "Darwin Centenary Programmes", BBC Memo, November 13, 1957, R34/981, BBC-WA.

[26] "Evolution: Nature's Technique of Creation", written and delivered by Gavin de Beer, June 29, 1958, BBC Home Service; and Christopher Holme, "Darwin Centenary", memo to P. H. Newby, May 8, 1957, R34/981, BBC-WA. All BBC copyright content reproduced courtesy of the British Broadcasting Corporation. All rights reserved.

[27] "Evolution and Religion", featuring Peter Medawar, David Newth and David Lack, July 2, 1958, BBC Third Programme; and P. H. Newby (Chief Assistant, Talks (Sound)), "Christianity and Natural Selection", BBC Memo, November 8, 1957, R34/981,

Table 8.1 All BBC broadcasts featuring significant content on Darwin broadcast across all BBC channels in 1958–1959

Year	*Channel*	*Series/Episode*	*Number of episodes*
1958	Third Programme	The Guillemot in Evolution	2
	BBC TV	Five Hundred Million Years	6
	Home Service	Bird's in Britain: Darwin's Finches	1
	Home Service	Evolution: Nature's Technique of Creation	1
	Home Service	Religion and Evolution	1
	Third Programme	Evolution in Action	4
	Network Three	Christian Outlook: Christian Belief and Natural Evolution	1
	Home Service	In Search of the Beagle	1
	Third Programme	Research: Hereditary and Evolution	1
	Home Service	The Naturalist: Animal Senses	1/5
1959	Third Programme	The Control of Evolution in Man	1
	Third Programme	Natural Selection Re-Examined	1
	Third Programme	The Real Origin of Species	1
	Home Service	The Naturalist: The Origin of Species	1
	Home Service	The Reith Lectures: The Future of Man	6
	BBC TV	Faraway Look	7
	Home Service	The Antiquity of Man	1
	Home Service	Two of a Kind: Centenarians	1
	Third Programme	The Darwinian Deception	1

List compiled using BBC Genome, repeats excluded

inspired by the ornithologist David Lack's book, *Evolutionary Theory and Christian Belief* (1957). More well-known for his research on Darwin's finches and speciation, in this book—also timed to capitalise on the Darwin Centenary—Lack aimed to resurrect discussion on religion and natural selection, arguing that characteristics such as humanity's self-awareness and free will were likely acquired by supernatural intercession. During discussion, Lack was politely taken to task by Medawar and the other scientist featured on the broadcast, zoologist David Newth, for suggesting *Homo sapiens'* unique characteristics and his proposed supernatural intervention, were beyond the scope of scientific study. Drawing on a similar argument to that made in his *Reith Lectures*, Medawar contended that while humanity's moral awareness may be beyond the purview of biological evolution, it was merely a social phenomenon, transmitted through non-genetic hereditary traditions, and like any social institution it could be studied rationally without recourse to spiritual arguments.[28]

The flagship of the anniversary broadcasts was the first major television treatment on evolution, *Five Hundred Million Years* (Chap. 3). Advertised with the tagline "[h]ow evolution looks today a hundred years after Darwin", the series featured talking head interviews with many of the scientists who also appeared on BBC radio's more intellectual centenary content.[29] Despite its idiosyncrasies, such as the studio debate format of the final episode, the series was an integral step in transferring the neo-Darwinian narrative already popular on radio science content to the small screen. The series went on to influence the format of BBC TV science content, particularly with regard to the narrative structures utilised for content dealing with the natural world (Chap. 6). The other major contribution to celebrations on BBC Television was the second series of *Faraway Look*, broadcast in September 1959 to coincide with the 125th anniversary of Darwin's visit to the Galapagos Islands aboard the *Beagle*.[30] *Faraway Look* was a spin-off from Peter Scott's (1909–1989) studio-based natural history show *Look* (1955–1968). The show built on the success of *Zoo Quest* and followed the naturalist and son of Antarctic explorer, Robert Falcon Scott, as he travelled to exotic locations filming the local fauna.[31] In a charming but somewhat amateur tradition, the 1959 series followed Scott, his wife Philippa (1918–2010) and camera operator Tony Soper (b.1929) as they sailed to the Galapagos to film the unique animals that inspired Darwin.[32] The expedition was part-funded by the International

[28] Peter Medawar, "Evolution and Religion", July 2, 1958, BBC Third Programme, 5–8. Transcript on microfilm at BBC-WA.

[29] "Five Hundred Million Years", *Radio Times*, April 18, 1958, 15.

[30] Peter Scott, "Our Galapagos Expedition", *Radio Times*, September 11, 1959, 3.

[31] Jean-Baptiste Gouyon, *BBC Wildlife Documentaries in the Age of Attenborough* (Springer Nature, 2019), 84.

[32] For more on how Scott's approach to natural history broadcasting increasingly became viewed as amateur by the professionalising team at BBC West England, which officially

Union for the Conservation of Nature under a project chaired by Julian Huxley to establish a research station on the islands.[33] In addition to four episodes on the Galapagos, the series began with three episodes filmed en route in the Virgin Islands, at the New York Zoological Society's research station in Trinidad and at the Smithsonian Institution's research station on the island of Barro Colorado in the Panama Canal.[34] In keeping with most natural history television of the period, the sequences filmed in the field focused on capturing animal behaviour and not on the species' biology or their evolution. As with Darwin's writing on the Galapagos, the series focused on species that had unique attributes influenced by their evolutionary isolation. The series was part of mid-twentieth-century developments that popularised the idea that Darwin discovered evolution in the Galapagos, an idea primarily driven by a coalition of evolutionary synthesis biologists and conservationists with Julian Huxley at their centre.[35] While the anniversary had been the inspiration for the series, substantive links to Darwin and his ideas were limited to wrap-around elements. This was typified by a post-hoc conversation with Julian Huxley inserted into episode four, in which he gave an overview of how the species found on the Galapagos had influenced Darwin's work.[36]

Much like the Chicago and London conferences, the predominant focus of the Darwin Centenary broadcasts was on showing how evolution was now the explanatory framework that unified the field of biology, based on Darwin's foundations. In addition to this promotion of the modern

became the Natural History Unit in 1957, see Gouyon, 99–102. As well as being a broadcaster Soper was a noted naturalist, publishing many books, including with fellow NHU Producer John Sparks (Chap. 6). See, for example, John Sparks and Tony Soper, *Penguins* (David & Charles, 1967).

[33] For more on the establishment of the research station and wider conservation efforts in the Galapagos, particularly those driven by Julian Huxley, see Elizabeth Hennessy, 'Mythologizing Darwin's Islands', in *Darwin, Darwinism and Conservation in the Galapagos Islands: The Legacy of Darwin and Its New Applications*, ed. Diego Quiroga and Ana Sevilla, Social and Ecological Interactions in the Galapagos Islands (Springer International Publishing, 2017), 65–90, https://doi.org/10.1007/978-3-319-34052-4_5.

[34] The International Union for the Conservation of Nature was established in 1948 under the impetus of Julian Huxley in his role as the first Director-General of the United Nations Educational, Scientific and Cultural Organization (UNESCO). Peter Scott, "Our Galapagos Expedition", 3.

[35] Hennessy, 'Mythologizing Darwin's Islands'; Elizabeth Hennessy, *On the Backs of Tortoises: Darwin, the Galapagos, and the Fate of an Evolutionary Eden* (Yale University Press, 2019), 116–146.

[36] See also the subsequent publication of the voyage, which featured many of Philippa Scott's still photographs: Peter and Philippa Scott, *Faraway Look II* (Cassell, 1960).

evolutionary synthesis, other content took "the popular approach" and focused on celebrating Darwin and his ideas as distinctly British turning points in history.[37] More critical or revisionist accounts of Darwin and his ideas were afforded only a very limited amount of airtime. Of the two notable critical accounts, the first, *Natural Selection Re-Examined*, a radio talk by the botanist Ronald Good (1896–1922), questioned the hegemonic application of natural selection, particularly in regards to emerging knowledge of certain plant species' phylogeny.[38] In the second, *The Darwinian Deception*, based on new research on Darwin's correspondence and notebooks, the historian of science Alistair Crombie (1915–1996) attempted to revise the popular understanding of Darwin as the archetypal Baconian scientist who meticulously compiled facts before formulating his theory.[39] However, these two critical episodes, one focused on science and one on the history of science, were vastly outnumbered in the schedule by more doting accounts. Unlike content on Darwin introduced in earlier chapters, and unsurprisingly given the celebratory nature of anniversaries, controversy was not a dominant framing used in the anniversary broadcasts during 1958–1959.

The broadcasts from 1958 to 1959 were part of a long process of reducing and mythologising Darwin's life and work, which during the first half of the twentieth century saw Darwin co-opted as the figurehead of attempts to unify disparate areas of biological study.[40] While not a central feature of most centenary broadcasts, the popular narrative of Darwin's life and work that persists today almost always features the controversy

[37] G. G. Mosley (Head of Overseas Talks and Features), "Darwin Centenary", BBC Memo, February 28, 1958, R34/981, BBC-WA. All BBC copyright content reproduced courtesy of the British Broadcasting Corporation. All rights reserved.

[38] "Natural Selection Re-Examined", written and delivered by Ronald Good, March 22, 1959, BBC Third Programme. For more on the academic argument that formed the basis of Good's radio lecture, see: Ronald Good, *Features of Evolution in the Flowering Plants* (Longmans, Green & Co., 1956).

[39] "The Darwinian Reception", written and delivered by Alistair Crombie, November 24, 1959, BBC Third Programme.

[40] Smocovitis, *Unifying Biology*; Hennessy, *On the Backs of Tortoises*; and Peter J. Bowler, 'The Specter of Darwinism: The Popular Image of Darwinism in Early Twentieth-Century Britain', in *Darwinian Heresies*, ed. Abigail Lustig, Robert J. Richards, and Michael Ruse (Cambridge University Press, 2004), 48–68. For more on Darwin's public profile during his own lifetime see: Janet Browne, 'Charles Darwin as a Celebrity', *Science in Context* 16, no. 1–2 (March 2003): 175–194, https://doi.org/10.1017/S0269889703000772.

that his ideas caused in Victorian society. Evidence of how this popular historical narrative fermented during the mid-twentieth century can be seen in the somewhat paradoxical role that Alfred Russel Wallace played in the centenary celebrations. On the one hand, the British scientific elite decided that the primary focus of celebrations should coincide with the centenary of Darwin and Wallace's 1858 papers, not the 1859 publication of Darwin's *Origin*. On the other hand, where Wallace did feature, in both broadcasts and the wider celebrations, he was at best a marginal figure, usually mentioned in passing in historical reflections that recounted events of 1858.[41]

In addition to the broadcasts listed in Table 8.1, the BBC also used the occasion to promote and celebrate Darwin around the world via their overseas services.[42] As part of this General Overseas Service content, Julian Huxley produced a short 15-minute talk on *Origin* as a turning point in the history of human thought. Huxley created the talk at the behest of the BBC German Service, which celebrated the Darwin anniversary at length under their "Projection of Britain" strand of broadcasting.[43] This form of soft diplomacy, which like some of the domestic content, celebrated Darwin to promote British science and ideals, was fast becoming a key component of the burgeoning "Darwin industry". As we shall see, this component became central in future Darwin anniversaries, particularly the 2009 bicentenary of Darwin's birth.[44]

[41] While internally some BBC producers referred to proceedings as the Darwin/Wallace Centenary, the overwhelming majority of correspondence called it the Darwin Centenary from the outset.

[42] As well as content broadcast on the BBC General Overseas Services in both English and local regional languages, there were also broadcasts on Darwin and evolution produced by broadcasters all over the world during 1958/1959, with several prominent radio and television broadcasts following the Chicago conference in the US. Sol Tax and Charles Callendar, *Evolution after Darwin*, Volume III, 263–270; and Smocovitis, 'The 1959 Darwin Centennial Celebration in America', 312–313.

[43] G. G. Mosley (Head of Overseas Talks and Features), "Darwin Centenary", BBC Memo, February 28, 1958, R34/981, BBC-WA; Lindley Fraser (Head of German Services), letter to Julian Huxley, January 29, 1958, 'Talks. Sir Julian Huxley, File III, 1950–1962', Rcont1, BBC-WA.

[44] John van Wyhe, 'Darwin Online and the Evolution of the Darwin Industry', *History of Science* 47:4 (1 December 2009): 459–473, https://doi.org/10.1177/007327530904700407; Michael Ruse, 'The Darwin Industry: A Guide', *Victorian Studies* 39, no. 2 (1996): 217–235.

Refining the Mythic Narrative of Darwin's Life and Influence

We can gain further insight from the BBC Darwin celebrations of 1958–1959 by considering them within the wider context of anniversaries at the public service broadcaster. With a growing number of channels across radio and television, the BBC regularly reviewed upcoming anniversaries that might be worthy of commemoration on radio or television. In the 1950s, BBC staff considered a wide range of future anniversaries, but few received more than one or two dedicated shows, if any at all.[45] By looking at who was shortlisted, who received dedicated broadcasts and who received no attention at all, we can learn about the priorities of public service broadcasters, and in turn the wider British establishment. Records show decisions were often contingent, related to other pressures on the schedule, the expected audience a historical figure might attract, and the connection to Britain and its empire.[46] Where anniversaries of scientists and their publications were concerned, the contemporary relevance and salience of their work was also a major influence. While Darwin received over 20 dedicated broadcasts in 1958–1959, the 300th birthday of another major figurehead of British science, Isaac Newton (1642–1726/1727) in 1942 only generated one 15-minute radio talk.[47] Following talks with the Royal Society, the BBC sought to commemorate the bicentenary of Jean-Baptiste Lamarck in 1944 with a talk by an esteemed expert.[48] Several academics advised that they saw no need, other than for geopolitical purposes to commemorate Lamarck, as one BBC producer summarised:

[45] For example, "Special Anniversaries", *BBC News Information Service*, September 1957, R34/981, BBC-WA.

[46] Other non-scientist anniversaries commemorated in 1958–1959 included a talk on the Cromwell Tercentenary by C. V. Wedgwood ("Servant of the Lord", November 10, 1958, BBC Home Service) and a talk on the 150th Anniversary of the birth of Abraham Lincoln by Columbia University Professor of History, David Donald (February 12, 1959, BBC Home Service). BBC, *Annual Report and Accounts for the year 1958–59* (HMSO, 1959), 42.

[47] "Sir Isaac Newton", written and delivered by James Jeans, December 25, 1942, BBC Home Service.

[48] Other significant anniversaries marked with a radio talk in the period that Lamarck was overlooked, include Isaac Newton (1942), Joseph Banks (1943), Antoine Lavoisier (1943), Nicolaus Copernicus (1943), Andreas Versalius (1943), and Louis Pasteur (1945). G. R. Barnes (Director of Talks), "Science Talks", December 14, 1943, R51/523/4, BBC-WA.

> [T]he best which can be said for Lamarck is to picture him as a man of great grasp and mental power, who profited by someone else's flash of intuition, and by the historical situation, and who, as a taxonomist, brought much order out of chaos; but Buffon before him on one side, and Erasmus Darwin on the other, are really names which remain much more significant.[49]

Subsequently, not only was a talk for Lamarck's bicentenary dropped from the domestic schedule, but it was also dropped from the BBC French service, despite being considered an important political gesture, as the French "still exalt Lamarck".[50]

It is clear that in commemorating Darwin, something on a scale not seen for other scientists occurred. While, given recent developments in epigenetics, Lamarck's reputation has been re-evaluated his influence remains marginal.[51] Yet study and celebration of Darwin has continued to grow, routinely eclipsing his co-author Wallace and other scientific luminaries such as Newton.[52] While there are myriad influences contributing to this wider canonisation, for broadcasters—at least initially—the popularity of broadcasts on Darwin and evolution led to a growth in content following the centenary. Extending the celebrations to commemorate the centenary of the infamous clash between T. H. Huxley and Bishop Samuel Wilberforce, BBC TV aired an episode of the religion department show *Meeting Point* (1956–1964) in July 1960. "The Battle of Oxford" featured a familiar cast, the Canon Charles Raven (Chap. 3), the naturalist James Fisher and T. H. Huxley's grandson Julian. The short discussion reinforced a mythic version of the infamous debate, drawing directly on *The Life and Letters of Thomas Henry Huxley* written by Julian's father, Leonard Huxley.[53]

Along with a very small handful of other iconic British historical figures, BBC producers seeking good audience figures continued to use

[49] Geoffrey Grigson, "Bi-Centenary of Lamarck: Professor R. C. Punnett", BBC Memo to Vincent Alford, June 14, 1944, R51/523/4, BBC-WA.

[50] Vincent Alford, "Bi-Centenary of Lamarck", BBC Memo to Geoffrey Grigson, May 23, 1944; and Vincent Alford, "Bi-Centenary of the Birth of Lamarck: August 1744", BBC Memo to Director of Talks, July 4, 1944, R51/523/4, BBC-WA.

[51] David Penny, 'Epigenetics, Darwin, and Lamarck', *Genome Biology and Evolution* 7:6 (2015): 1758–1760, https://doi.org/10.1093/gbe/evv107.

[52] John C. Greene, 'Reflections on the Progress of Darwin Studies', *Journal of the History of Biology* 8:2 (1975): 243–273.

[53] *Meeting Point*, "The Battle of Oxford", July 31, 1960, BBC TV; Leonard Huxley, *The Life and Letters of Thomas Henry Huxley* (D. Appleton & Co., 1901), 192–199.

various, sometimes tenuous, Darwin-related anniversaries and milestones as a hook for science and history content.[54] As with the 1958–1959 broadcasts, when viewed as historical artefacts we can see how these shows always refracted and reimagined Darwin through the lens of contemporary society. For example, take the listings for a short radio remembrance, "Darwin at Westminster" broadcast in April 1982 to coincide with the centenary of Darwin's death:

> In the light of the religious controversies that had surrounded his work, how appropriate was his final resting place? How suitable is it now? As evolutionary theory comes under attack from creationists on the right and radicals on the left, is Darwin's reputation as the greatest of British naturalists in decline? Will it, too, be buried beneath the sacred slabs?[55]

Nearly all of the content produced for the centenary of Darwin's death in 1982 was explicitly framed through the lens of US scientific creationism (Chap. 7). However, connecting anniversaries to current topical issues is not the only device used by broadcasters to attract an audience. Another common trope, especially in factual historical broadcasting, is the "forgotten figure".

Also from the 1982 centenary content, take for example, the television drama "The Forgotten Voyage", written by one of the popularisers of the aquatic ape hypothesis, Elaine Morgan (Chap. 7). This film-length drama charted the life and work of Alfred Russel Wallace. A curious combination of dramatised dialogue, natural history film and travelogue, the docu-drama used the written historical record, in particular Wallace's own diaries, to narrate and bring to life his travels in Southeast Asia. Although the

[54] While across the twentieth-century BBC broadcast content on Darwin easily eclipses any other historical or contemporary scientist, historical figures from areas such as politics, royalty and cinema have had much more airtime dedicated to their lives. Notable examples include Winston Churchill, Queen Victoria and Charlie Chaplin. For more on the contemporary role that celebrity plays in the public understanding of evolution see Amy Unsworth and David Voas, 'The Dawkins Effect? Celebrity Scientists, (Non)Religious Publics and Changed Attitudes to Evolution', *Public Understanding of Science*, 25 (2021), 0963662521989513, https://doi.org/10.1177/0963662521989513.

[55] "Darwin at Westminster", written and presented by James Moore, April 18, 1982, BBC Radio 4, *The Radio Times*, April 15, 1982, 39. Other content during 1982 that used this creationist frame included an episode of the BBC Radio 1 youth debate show *Talkabout* (1979–1983) in which members of a youth club in Sheffield debated evolution versus creation. *Talkabout*, "The Apes v The Angels", April 6, 1982, BBC Radio 1.

Fig. 8.1 The curator at the British Museum introducing a young Alfred Russel Wallace (Tim Preece) to Charles Darwin (Tony Steedman) in a scene from the docudrama "The Forgotten Voyage" (*The World About Us*, BBC Two); and inset the front cover of the *Radio Times* promoting the film (*Radio Times*, December 16, 1982)

docudrama was pitched as reintroducing the public to this forgotten figure from the history of British science, Wallace was situated as just one man within a wider network of British gentleman studying natural history across the British Empire. In an early scene, a young Wallace was awestruck when introduced to Charles Darwin among the drawers of insects at the British Museum (Fig. 8.1).[56] Reflecting the fact that the episode was commissioned as part of the natural history series *The World About Us*, which had been established by David Attenborough in 1967, many of the sequences were filmed on location.[57] In a crowded Christmas schedule,

[56] *The World About Us*, "The Forgotten Voyage", written by Elaine Morgan, produced and directed by Peter Crawford. December 24, 1982, BBC Two. Interested readers can watch the whole docudrama online: https://www.youtube.com/watch?v=Z1eQ6DadodA *(last accessed 12/06/21)*.

[57] Although more wide-ranging and often anthropologically focused in its content than other natural history series in the period, despite the odd historical re-enactment, the format for *The World About Us* was almost always a traditional natural history documentary. Combining BBC commissioned episodes with syndicated films made by international part-

"The Forgotten Voyage" was well received, with its success leading to another dramatised special on the history of evolutionary science, "The Garden of Inheritance" (1984), in which the same team of Elaine Morgan and Peter Crawford adapted the life story of Gregor Mendel.[58]

We must remember that this attempt to remember the forgotten figure of Wallace was just one broadcast among a larger number produced specifically to commemorate the centenary of Darwin's death. Just a few months before "The Forgotten Voyage", the BBC had aired the biopic Hollywood film *The Darwin Adventure* (1972).[59] And it was Darwin's own voyage on the *Beagle*, which had originally proven the popularity of the historical drama format, first on radio, with Orwell's episode in the *Voyages of Discovery* (1946) series (Chap. 5), and then on television with a seven part historical docudrama, *The Voyage of Charles Darwin* (1978).[60]

While over the decades the voyage of the *Beagle* remained a persistent feature of broadcasts on Darwin, the framing used in broadcasts over the second half of the twentieth century shifted depending on popular trends and the target audience. By looking at the long-running science series *Horizon*, we can neatly see how the changing media communication of evolution and its relationship with Charles Darwin's original research evolved. Of approximately 1200 episodes of *Horizon* broadcast between 1964 and 2009, over 35 contained substantive content on evolutionary science.[61] If we focus on those that featured Darwin in their titles, the

ners, such as *National Geographic*, early seasons included famous naturalists, such as the French marine conservationist and filmmaker Jacques Cousteau, the primatologist Jane Goodall, and a Canadian Broadcasting Corporation episode on Darwin. David Attenborough, *Life on Air: Memoirs of a Broadcaster* (Princeton University Press, 2002), 211–212; *The World About Us*, "The Undersea World of Jacques Cousteau", November 3, 1968, BBC Two; *The World About Us*, "The Miss Goodall and the Wild Chimpanzees", May 26, 1968, BBC Two; and *The World About Us*, "Darwin", September 1, 1968, BBC Two.

[58] Chris Dunkley, "Few Goodies at Christmas", *The Financial Times*, January 5, 1983, 13; and *The Natural World*, "The Garden of Inheritance", written by Elaine Morgan, produced and directed by Peter Crawford, February 6, 1984, BBC Two.

[59] *Saturday Cinema: The Darwin Adventure*, October 30, 1982, BBC Two.

[60] "The Voyage of the 'Beagle'", *Voyages of Discovery*, produced by Rayner Heppenstall, written by George Orwell, BBC Home Service, March 29, 1946; and *The Voyage of Charles Darwin*, written by Robert Reid, produced by Christopher Ralling and Ned Kelly, and directed by Martyn Friend, October to December 1978, BBC Two.

[61] This list of 36 episodes of *Horizon* that have featured substantive content on evolutionary science is not exhaustive, it covers the period 1964–2009, and many episodes on related subjects such as embryology, genetic modification and palaeontology have not been included. A full list of all *Horizon* episodes can be found online: https://en.wikipedia.org/wiki/List_

episodes read like the narrative arc of a Hollywood film franchise (Table 8.2). In "Darwin's Bulldog" (1971) the mythic narrative of the Huxley-Wilberforce debate, as featured earlier in "The Battle of Oxford" (1960), got the historical dramatisation treatment in a combative conflict-centred narrative, which proposed that scientific discovery required more than lone figures and reproducible truths to become accepted knowledge. In 1977's "Darwin's Dream", the docudrama approach was this time used so that a time-travelling Darwin could explore the latest neo-Darwinian science built on his original ideas. In addition to cheering the progress made—"it's becoming clearer!"—the resurrected Darwin would have been pleased to hear contemporary scientists still using the language of awe and wonder.[62] This was Darwin as grandfather of a progressive materialistic scientific project. By 1981, Darwin was purportedly in need of defence again. In "Did Darwin get it Wrong?" producers drew an audience by advertising that Darwin was again under attack from evolution's old adversaries, religious fundamentalists. However, as we saw in Chap. 7, the episode focused more on new theories challenging contemporary understanding, such as Niles Eldredge's (b.1943) work on punctuated equilibria, all of which was within the remit of evolutionary science and normal scientific practice (Table 8.2).[63]

Finally, by 1998, Darwin's position as an icon in the history of science was seemingly immovable. The first part of a BBC/OU special for "Evolution Weekend" gave a now-familiar narrative of Darwin's life,

of_Horizon_episodes *(last accessed 15/03/2021)*. "Professor J.B.S. Haldane, Obituary" (1964), "The Living Stream" (1965), "Genes in Action" (1966), "Comfort on Ageing" (1968), "The Manhunters" (1970), "Darwin's Bulldog" (1971), "The Fierce People" (1971), "What Is Race?" (1972), "Search for Life" (1974), "The Race for the Double Helix" (1974), "The Lysenko Affair" (1974), "The Selfish Gene" (1976), "The Ape that Stood Up" (1977), "One of Nature's Hotels" (1977), "Genetic Roulette" (1977), "Darwin's Dream" (1977), "Bags of Life" (1978), "Did Darwin Get it Wrong?" (1981), "Tropical Time Machine" (1983), "Professor Bonner and the Slime Moulds" (1984), "The Blind Watchmaker" (1987), "Chimp Talk" (1993), "Life Is Impossible" (1993), "The Predator" (1994), "Tibet, The Ice Mother" (1995), "Darwin the Legacy" (1998), "Designer Babies" (1999), "Wings of Angels" (1999), "The Missing Link" (2001), "The Ape That Took Over the World" (2001), "Cloning the First Human" (2001), "The Day We Learned to Think" (2003), "Neanderthal" (2005), "A War on Science" (2006), "My Pet Dinosaur" (2007), and "What's the Problem with Nudity" (2009).

[62] Nancy Banks-Smith, "Darwin's Dream", *The Guardian*, September 24, 1977, 10.

[63] *Horizon*, "Did Darwin get it Wrong?" Written and produced by Alec Nisbett, narrated by Paul Vaughan, 30 March 1981, BBC 2.

Table 8.2 The *Radio Times* listings for the episodes of *Horizon* that featured Darwin as a central component

Date	*Episode*	*TV listing*
1971	Darwin's Bulldog	T. H. Huxley and the fight for Darwinism. "In so proudly claiming descent from monkey, ape or baboon, does he do so on his grandfather's side or his grandmother's?" This now-legendary challenge was thrown to T. H. Huxley by the Bishop of Oxford in 1860 and reduced a British Association meeting to chaos. The dramatic events leading up to this remark are re-enacted in tonight's Horizon about Thomas Henry Huxley, father of the famous Huxley family, originator of the term "agnostic", man of science, educator and the most famous protagonist of "The Origin of Species". Without Huxley's ardent and brilliant campaigning, it is conceivable that Victorian society would have ignored Darwin's theory of evolution and rejected the most significant idea of the nineteenth century.
1977	Darwin's Dream	Darwin's theory of evolution transformed our view of the world. But what would he think of the progress we have made since? Tonight, Darwin is reincarnated—in the form of Lee Montague—and visits various laboratories in modern Britain to ask searching questions of the scientists themselves. He does not know what a gene or a chromosome is-let alone viruses, enzymes and DNA. As a result the scientists have to explain with clarity and ingenuity how the beautiful, fascinating, organised chaos of life at the molecular level works. Darwin discovers that, in spite of enormous advances which have brought us to the brink of being able to manipulate evolution ourselves, we do not yet fully understand the rules which govern the harmony of the natural world.
1981	Did Darwin get it Wrong?	How did the elephant get its trunk? Or the giraffe its neck? Darwin's theory of evolution says that societies evolve slowly from one form to another. So ancient fossils and bones should include an elephant with half a trunk, or seven-eighths of a giraffe's neck. But they don't. And these gaps occur in all classes of animal fossils, from molluscs to man. So the established theory is now under strong attack both by Biblical Creationists, who offer well-argued objections to Darwin, and by a range of scientists who offer an alternative explanation. Could evolution occur in a series of jerks, switching suddenly into new creatures? With the help of alligators and beetles, chimpanzees and dinosaurs, elephants and fruit-flies, Horizon examines the evidence.

(*continued*)

Table 8.2 (continued)

Date	*Episode*	*TV listing*
1998	Darwin: The Life/ Darwin: The Legacy	Two-part special for Evolution Weekend. Part I: This documentary examines the paradox of Charles Darwin, a private and conventional man who developed one of the most revolutionary ideas in science. Part II: A film that explores the changing impact on society of Darwin's ideas, from those who first used them to justify extreme capitalism to supporters of eugenics. Leading voices in modern evolutionary biology, however, feel that they have wider implications, recalling Darwin's original message: that human nature is part of a larger picture.

List compiled using BBC Genome

explaining, "how the man who killed God came to receive his country's highest blessing, laid to rest in Westminster Abbey with the immortals of his nation's history".[64] In the second part, "Darwin: The Legacy", the focus shifted to wider societal impacts of Darwin's work. Here again was the progressive arc, past projects justified in the name of evolution, such as eugenics, were morally bankrupt, while contemporary calls about what evolutionary science means for society today were a call to arms to positively change human behaviour.[65] After an overview of recent popular

[64] "Darwin: The Life", March 28, 1998, BBC Two, 03:50–04:10: https://archive.org/details/BBCDarwinTheLifeMar1998/BBC+Darwin+-+The+Life+(Mar+1998).VOB *(last accessed 15/06/21)*. The episode drew heavily on the works of historians of science and Darwin biographers, in particular Adrian Desmond, James Moore and Robert M. Young. In the opening talking head sequences, Moore and Desmond both pose holding their biography *Darwin: The Life of a Tormented Evolutionist* (Penguin, 1992).

[65] The documentary featured a clip of Julian Huxley in a film produced by the Eugenics Society called *Heredity in Man* (1937) in which Huxley extolled the virtues of population control via restricting the reproduction of "defectives", saying it would be better if disabled people were not born. The full 14-minute film can be watched online: https://player.bfi.org.uk/free/film/watch-heredity-in-man-1937-online *(last accessed 15/06/21)*. While following the atrocities of the Second World War, Huxley's position on eugenics would alter significantly, he remained a supporter of a reformed type of eugenics that operated via social planning; both modes of his thinking on this subject were inextricably linked to his vision for evolutionary humanism. Paul Weindling, '"Julian Huxley and the Continuity of Eugenics in Twentieth-Century Britain"', *Journal of Modern European History* 10:4 (2012): 480–499, https://doi.org/10.17104/1611-8944_2012_4.

developments, including punctuated equilibria, the selfish gene, E. O. Wilson's socio-biology and evolutionary psychology, the episode zoomed out to outline how evolution is still political. Again refracting contemporary socio-political concerns, in particular those of the growing environmental movement, the episode finished by introducing research from the evolutionary sciences that showed humanity was now the key vector disrupting global ecosystems.[66]

Across all these episodes of *Horizon*, whether being used to defend science from resurgent creationists or as a clarion call for addressing the ensuing ecological crisis, broadcasters used Darwin and Darwinism in a totemic mythic mode. The historical flashback, first developed for educational radio broadcasts by Rhoda Power in the early 1940s (Chap. 4), and developed further in early radio dramas (Chap. 5), was by the 1970s an expected part of the format for TV science broadcasts. As with Orwell's radio drama, these flashbacks frequently centred on conflict were used to highlight progressive developments since Darwin and extolled the successes of the scientific project more generally.

Broadcasting the 2009 Darwin Anniversary, a Reflective Coda

Following the trajectory of the previous decades, the 2009 Darwin Anniversary was once more celebrated on an unprecedented scale. Again, the BBC was at the centre of celebrations in the UK and beyond as they used the occasion to celebrate Darwin as the most eminent and influential British success story. The BBC produced an astounding 70–100 episodes on national radio and television (Table 8.3), alongside a programme of regional broadcasts and events, and dedicated materials for schools and education via their partnership with the Open University.[67] With more channels now operating in the UK, broadcasts also aired on Channel 4, cable channels such as The History Channel and online streaming

[66] "Darwin: The Legacy", March 29, 1998, BBC Two. Available online: https://archive.org/details/BBCDarwinTheLifeMar1998/BBC+Darwin+-+The+Legacy+(Mar+1998).VOB *(last accessed 15/03/21)*.

[67] An archived website giving an overview of all of the BBC activities made as part of "Darwin Season 2009" including links to much of the BBC Regional events and coverage can be found here: "Darwin: The Genius of Evolution", BBC Website (Archived), http://www.bbc.co.uk/darwin/ *(last accessed 16/06/21)*.

Table 8.3 A selection of the major BBC broadcasts featuring significant content on Darwin broadcast across all BBC channels in 2008–2009

Year	*Channel*	*Series/episode*	*No. of episodes*
2008	BBC One	Life in Cold Blood	5
	Radio 4	Hunting the Beagle	5
	BBC HD	Galapagos	1/3
	BBC Two	The Supersizers ... Go Victorian	1/6
	BBC Four	Inherit the Wind	1
	Radio 3	BBC Proms Feature: In the Beginning Was the Song	1
2009	Radio 4	Darwin: In Our Time	4
	Radio 4	Dear Darwin	5
	Radio 4	Great Lives: Alfred Russel Wallace	1/9
	BBC Four	What Darwin Didn't Know	1
	Radio 4	Darwin: My Ancestor	4
	BBC One	Charles Darwin and the Tree of Life	1
	BBC Four	Darwin's Struggle: the Evolution of the Origin of the Species	1
	Radio 3	The Essay: Darwin's Children/Darwin's Descendants	5
	Radio 4	Sunday Service: 200th anniversary of Charles Darwin's birth	1
	Radio 3	Sunday Feature: Darwin's Conundrum	1
	Radio 3	Words and Music: The Ascent of Man	1
	Radio 4	The Material World: Darwin and the Eye	1
	BBC Two	Jimmy Doherty in Darwin's Garden	3
	BBC Two	Darwin's Dangerous Idea	3
	BBC Two	Did Darwin Kill God?	1
	Radio 4	The Darwinian Sistine Chapel	1
	Radio 2	Mike Harding: Darwin Song Project	1
	BBC One	Uncovering Our Earliest Ancestor: The Link	1
	Radio 4	The Material World: Cheltenham Science Festival	1
	Radio 4	The Eureka Years: 1866	1/4
	Radio 4	Leading Edge: Attitudes to Darwin	1/9
	Radio 4	Something Understood: The Tree of Life	1
	Radio 3	The Origins of the Origin	1
	Radio 4	The Saturday Play: Journey into Space—The Host	1
	Radio 4	A Life with ... Microbes	1
	BBC Two	Classic Goldie	2
	BBC One	Life	10
	Radio 4	Aping Evolution	2
	Radio 4	The Saturday Play: The Great Tennessee Monkey Trial	1
	Radio 4	Leading Edge: On the Origin of the Species	1

List compiled using BBC Genome, excludes repeats, Schools/Learning Zone content, BBC World Service content and regular magazine shows that featured less than a whole episode on Darwin services.[68] As well as technological developments, like the proliferation of media content via the Internet, the explosion in the number of Darwin broadcasts in 2009 was influenced by several other factors, such as the further professionalisation of science communication practice.[69] Globally, events and broadcasts were so numerous and diffuse that to study them in-depth could easily fill a whole book itself.[70] In this closing section, rather than comprehensive coverage of the 2009 Darwin Anniversary, I highlight the major features of the BBC "Darwin Season 2009", and then use specific broadcasts from the celebrations as a vantage point to review and summarise the developments outlined in the previous chapters. This approach shows how contingent decisions, technological innovations and individual ideologies combined over the course of the twentieth century to shape the very formats of broadcasts on evolution in the UK (Table 8.3).

As with the 1958–1959 celebrations, most BBC departments contributed content for the 2009 anniversary. However this time, efforts were more centrally co-ordinated, with broadcasts strategically placed in the

[68] Following the BBC, Channel 4 had the next most Darwin content during 2008/2009, while the other main terrestrial broadcaster ITV had very little dedicated content. For examples see: *The Genius of Charles Darwin*, directed by Russell Barnes and Dan Hillman, written and presented by Richard Dawkins, August 2008, Channel 4; "The Battle for Charles Darwin", History Channel, September 25, 2009; and "Darwin: 150 Years of Evolutionary Thinking", *Yale Peabody Museum of Natural History YouTube Channel*, https://www.youtube.com/watch?v=SunF2buXOUg *(last accessed 16/03/21)*.

[69] For more on the development of work on the public understanding of science in the UK see Jane Gregory and Simon Jay Lock, 'The Evolution of "Public Understanding of Science": Public Engagement as a Tool of Science Policy in the UK', *Sociology Compass* 2:4 (2008): 1252–1265, https://doi.org/10.1111/j.1751-9020.2008.00137.x. For an account that deals with the growth of celebrity with regard to elite science popularisers see Declan Fahy, *The New Celebrity Scientists: Out of the Lab and Into the Limelight* (Rowman & Littlefield, 2015).

[70] One of the most high-profile releases that coincided with the anniversary year was the film *Creation* (Recorded Picture Company/BBC Film, 2009), which debuted at the Toronto Film Festival in September 2009, before being distributed globally. Starring Paul Bettany and Jennifer Connelly as Charles and Emma Darwin, the biopic, which received a mixed reception from critics and popular audiences, deals with the challenges facing Darwin during the writing of *Origin*, in particular the effects on his personal faith in God and the grief he suffered following his daughter's death. "Darwin biopic to launch Toronto", *BBC News Online*, July 15, 2009, http://news.bbc.co.uk/1/hi/entertainment/8151456.stm *(last accessed 16/03/21)*; "Creation", Rotten Tomatoes Website, https://www.rottentomatoes.com/m/1205717-creation *(last accessed 16/03/21)*; and Rowan Hooper, "Review: *Creation: The Movie*", *New Scientist*, September 23, 2009, 49.

calendar, and content being co-ordinated across platforms.[71] The result was a remarkably diverse array of broadcasts under the "Darwin Season" banner (Table 8.3), with a concentration of programming in two spikes coinciding with the 200th anniversary of Darwin's birth in February 2009, and the 150th anniversary of *Origin*'s publication in November 2009. As well as major new commissions, the scheduling again saw Darwin used widely as an audience hook, with some rather tenuous attempts to link shows to the nineteenth-century scientist.[72] Unlike in 1958–1959, when the achievements of modern evolutionary biology and the evolutionary synthesis shaped the majority of broadcasts, the multi-department, multi-channel scale of 2009 saw Darwin's biography and his wider influence on contemporary society dominate proceedings.[73]

A wide-ranging cultural programme accompanied the main broadcasts, both at home and abroad, with the BBC collaborating with the British Council to cover their *Darwin Now* project.[74] While much of this content aired on the BBC World Service, Darwin in a global context also featured on domestic Radio 4 broadcasts.[75] This form of soft diplomacy, which

[71] Commissioning editor for factual television content, Martin Davidson acted as the commissioning editor for the season. "The BBC's Darwin Season: marking the life and work of Charles Darwin – introduction", *BBC Press Pack*, January 21, 2009, http://www.bbc.co.uk/pressoffice/pressreleases/stories/2009/01_january/21/darwin.shtml *(last accessed 16/03/21)*.

[72] For example, the pop station Radio 2 managed a tie-in in with an episode exploring the Shrewsbury Folk Festival's Darwin Song Project.

[73] As an illustration, around 70–100 broadcasts referred to Darwin in their *Radio Times* listing in 2008–2009 (Table 8.3), while as per the corpus introduced in Fig. 1.1, only 34 episodes (minus repeats) used the word evolution in a biological context during the same time period.

[74] The British Council is a charity and non-departmental public body sponsored by the British Foreign, Commonwealth and Development Office tasked with promoting the UK, via international cultural and educational activities. The *Darwin Now* project ran activities in over 50 countries worldwide including workshops, a website, a mobile exhibition and a major conference held at the Bibliotheca Alexandrina in Egypt in November 2009. Fern Elsdon-Baker, 'Creating Creationists: The Influence of "Issues Framing" on Our Understanding of Public Perceptions of Clash Narratives between Evolutionary Science and Belief', *Public Understanding of Science* 24:4 (2015): 422–439, https://doi.org/10.1177/0963662514563015; Amal Bakry, 'Cobranded Diplomacy: A Case Study of the British Council's Branding of "Darwin Now" in Egypt', *International Journal of Communication* 14:0 (2020): 20.

[75] The episode of *Leading Edge*, "Attitudes to Darwin" discussed creationism in US classrooms and a Darwin exhibition in Turkey, while the episode, "On the Origin of Species"

drew on Darwin to promote British science and cultural ideals, had grown greatly in scale and sophistication from its more limited deployment in 1958–1959.

With celebrations largely limited to 2009, focused mainly on Darwin's birthday and the November publication of *Origin*, Wallace was even more absent than he had been in 1958–1959.[76] Reflecting how much further popular narratives of the history of evolution had marginalised Wallace since the mid-twentieth century; he was the focus of just one short late night radio episode.[77] The lack of commemoration of the 150th anniversary of the discovery of natural selection, and Wallace's role in it, occurred not just in BBC coverage, but also across the majority of celebrations globally in 2009.[78] The erasure of Wallace was subsequently noted, and broadcasters used the opportunity to once again cast Wallace in the "forgotten figure" mould over the following years.[79]

Across content produced for the 2009 Darwin Season, we can see myriad hallmarks of the approaches, narrative styles and formats introduced over the preceding chapters. Like all major factual science content, from Huxley's earliest broadcasts onwards, celebrity biologists and popularisers dominated content on both evolutionary science and its history, most notably David Attenborough on the BBC, and Richard Dawkins on Channel 4.[80] Following earlier patterns, in-depth and high-level coverage

featured the *British Council* Darwin Now conference held in Egypt. Readers can listen to these two episodes online: *Leading Edge*, "Attitudes to Darwin", June 4, 2009, BBC Radio 4, https://www.bbc.co.uk/programmes/b00kq564; and *Leading Edge*, "On the Origin of the Species", November 26, 2009, BBC Radio 4, https://www.bbc.co.uk/programmes/b00nyz7n *(last accessed 16/06/21).*

[76] The sheer scale of BBC broadcasts planned for 2009 meant that during 2008 when the Wallace/Darwin anniversary occurred there were only a few repeats and one minor new series on evolution or Darwin across the whole year, an effect that is evidently seen in Fig. 1.1. The one new series was *Hunting the Beagle* written and presented by Dr Robert Prescott, April 2008, BBC Radio 4.

[77] *Great Lives*, presented by Matthew Parris and produced by Miles Warde, January 23, 2009, BBC Radio 4.

[78] George W. Beccaloni and Vincent S. Smith, "Celebrations for Darwin downplay Wallace's role", *Nature*, 451, 1050 (2008), https://doi.org/10.1038/4511050d.

[79] See, for example, *Bill Bailey's Jungle Hero*, presented by Bill Bailey, produced by Sam Hodgson and Tuppence Stone, October 2019, BBC Four. For more on the series, including short clips see https://www.bbc.co.uk/programmes/p0160nxk *(last accessed 21/06/2021).*

[80] For example, *Charles Darwin and the Tree of Life*, presented by David Attenborough, produced by Sacha Mirzoeff, February 1, 2009, BBC One; *The Genius of Charles Darwin*,

that featured more practicing contemporary scientists and historians was limited to radio and BBC Four, the BBC's new home for non-mainstream cultural and educational content.[81]

Like in 1958–1959, there were some fringe perspectives, but the 2009 season of broadcasts was almost completely devoid of programming in a critical mode. Again, this lack of critical programming may have been an effect of the celebratory frame of an anniversary. However, if we look back over the prior decade of shows on evolutionary themes, there are very few examples that presented new contested areas in the biological sciences, or revisionist historical accounts. Where new biological ideas were covered, such as a 2005 episode of *Horizon* on epigenetics called "The Ghost in Your Genes", discussion of evolution as an overarching narrative was minimal or completely absent.[82] Despite the celebratory agenda, some 2009 Darwin Season broadcasts still used a controversy-centred framing. Two of the major television contributions prominently featured danger, conflict and fear; the mini-series *Darwin's Dangerous Idea* (2009) and a one-off special *Did Darwin Kill God?* (2009).[83] Across the whole season of broadcasts, as with the dramatisations featured in the previous chapters, whether just flashbacks or whole series, historical controversies were used to highlight the progressive attitudes of contemporary British society.

While in 1958–1959 the content clearly reflected the pressing concerns of evolutionary scientists at the time, across the 2009 broadcasts it is harder to discern one overriding contemporary concern that shaped content. If a unifying theme can be found, it is a shallow emphasis on celebrating Darwin and his ideas. The influence of Darwin and evolution in

directed by Russell Barnes and Dan Hillman, written and presented by Richard Dawkins, August 2008, Channel 4.

[81] For example in the BBC Radio 4 series *Dear Darwin* contemporary biologists, Craig Venter, Jonathan Miller, Jerry Coyne, Peter Bentley and Baruch Blumberg, wrote letters to Darwin in which they outlined how his work had shaped their own research. A similar concept was used in the BBC Four one-off, *What Darwin Didn't Know*, which was written and presented by the biologist Armand Marie Leroi.

[82] *Horizon*, "The Ghost in Your Genes", produced by Nigel Paterson, November 3, 2005, BBC Two.

[83] Despite the provocative title and accompanying controversy-centred framing, in *Did Darwin Kill God?* the philosopher and theologian Conor Cunningham presented an accommodationist account in keeping with earlier BBC content on evolution and religion. *Darwin's Dangerous Idea*, produced and directed by Kate Bartlett, presented by Andrew Marr, March 2009, BBC Two; and *Did Darwin Kill God?*, produced by Emily Davis and Jean Claude Bragard, written and presented by Conor Cunningham, March 2009, BBC Two.

general on humanity was covered from a much wider range of subjects and perspectives than prior attempts. Some content was tenuous, both in its link to Darwin and even to broader evolutionary themes. For example, in *Classic Goldie*, the drum n bass DJ, Goldie, was set the challenge of composing a piece of classical music with the theme of "evolution".[84] In more considered reflections on Darwin's broader impact, such as *Darwin's Dangerous Idea*, in which the political journalist Andrew Marr covered Darwin's impact on religion and morality, political structures and our relationship with the natural world, there were many echoes with earlier approaches.

Given the amount of content generated in 2009, the most notable omission in the schedule was that there was no grand treatment of evolution. While there were successful contributions from the Natural History Unit (NHU), with many of the usual touchstones such as the importance of the Galapagos Islands, none of the broadcasts were explicit evolutionary epics. For example, although the final offering in the *Life* franchise, *Life in Cold Blood* (2008) had all the hallmarks of NHU series in the humanist blockbuster mode, its focus on reptiles and amphibians meant it largely lacked a broader progressive evolutionary narrative.[85] Likewise, wide ranging as the history of science content was, none was on the scale of prior humanist blockbusters, such as *Ascent*. To account for these seeming gaps in the 2009 anniversary programming we should first look to wider trends.

Over the course of the twentieth century, the popularity of universal narratives of science and human progress waxed and waned.[86] By the beginning of the twenty-first century, at least in British broadcasting, it would seem that explicit grand visions of man's evolutionary future, as expounded by Medawar at the start of this chapter or by Huxley in 1925 (Chap. 2), had fallen out of favour. As Zakariya identifies when comparing *Cosmos: A Personal Voyage* (1980) with its 2014 remake, *Cosmos: A Spacetime Odyssey*, scientific epics and cosmic accounts still abounded in

[84] The BBC Concert Orchestra and the London Philharmonic Orchestra performed the resulting piece of music, *Sine Tempore* as part of a BBC Proms concert comprised of music connected to Darwin. Their performance can be watched online: https://www.youtube.com/watch?v=axNCWpM1fQk *(last accessed 16/06/21)*.

[85] Patrick Foster, "Eighth Bafta slides into hands of Attenborough for reptile series", *The Times*, April 27, 2009, 9.

[86] Nasser Zakariya, *A Final Story: Science, Myth, and Beginnings* (University of Chicago Press, 2017).

the twenty-first century, they had just shifted in mode, to be more conscious of our limitations, reframing the hero from science alone to humanity as a whole. Less scientifically determinist in tone, they are confident of science's ability to solve problems, but also more acknowledging of multicultural perspectives and the huge failure of our ability, thus far, to limit the pressure our species is putting on global ecological systems.[87]

Yet observing these shifts in recent grand epic documentaries and situating them within wider trends on the popularity of universal narratives is only part of the story here. The seeming lack of prevalence of the evolutionary epic and humanist blockbusters across the 2009 broadcasts is also because this progressive evolutionary narrative is now the unquestioned norm. As we saw in the very opening of this book, by the early twenty-first century, the secular position that earlier exponents explicitly argued for, namely that humanity through knowledge provided by the biological sciences controlled its own future and fate, has become the implicit normative framing for broadcasts. In the interim chapters, we have seen how a myriad of departments, genres and individual writers, producers and broadcasters were responsible for the development of the visual devices and narrative tools that enabled this shift. We have seen how this evolutionary narrative and its popularisation was part of the emergence of a wider way of presenting popular science, rooted in scientific humanism, materialism and progressivism. Yet importantly, we have also seen that its adoption was by no means inevitable. Especially given in its earlier nineteenth-century incarnation, the evolutionary epic was often employed in support of religious worldviews.[88]

While broadcasting is only one component in the normalisation of this narrative form, I have shown that without the integral contributions of media producers, the visionary accounts of synthetic popularisers like Julian Huxley and Jacob Bronowski would not have become as ensconced in British society as they are today. Across the twentieth-century, British broadcasters acted as communicators, translators and amplifiers of ideas, narratives and beliefs that have played an important role in shaping British society. Throughout the book, I have not attempted to place any value judgement on the different approaches taken by broadcasters and communicators. Whether taking advantage of a newsworthy angle, a popular

[87] Zakariya, 409–417.

[88] For examples, see, Ian Hesketh, 'The Recurrence of the Evolutionary Epic', *Journal of the Philosophy of History* 9:2 (2015): 204–209.

narrative form or a university course on evolution, media producers created content in response to immediate pressures.

There is no correct way to present evolutionary themes, no inherent reason that we should favour the didactic over the rhetorical. Today at a societal level, whether we as a whole species are actually able to control our own biological future and fate is currently being tested by insufficient action in the face of numerous environmental crises. Whether Huxley's evolutionary humanism, or the multitude of other progressive iterations on what evolution says about our future I have introduced, is one part of the answer, remains to be seen. As a historian, I am much better at describing what has been, than proposing what might be, but if we can take one thing forward from the history of evolution in broadcasting as presented in this book, it is that it has frequently been more robust and more successful in its aims when plural and diverse approaches have been allowed to address the subject and its challenges.

References

Attenborough, David. *Life on Air: Memoirs of a Broadcaster*. Princeton University Press, 2002.

Bakry, Amal. 'Cobranded Diplomacy: A Case Study of the British Council's Branding of "Darwin Now" in Egypt'. *International Journal of Communication* 14:0 (2020): 20.

Banks-Smith, Nancy. "Darwin's Dream", *The Guardian*, September 24, 1977, 10.

BBC, *Annual Report and Accounts for the year 1958–59*. HMSO, 1959.

Beccaloni, George W., and Vincent S. Smith, "Celebrations for Darwin downplay Wallace's role", *Nature*, 451:1050 (2008), https://doi.org/10.1038/4511050d.

Bowler, Peter J. 'The Specter of Darwinism: The Popular Image of Darwinism in Early Twentieth-Century Britain'. In *Darwinian Heresies*, edited by Abigail Lustig, Robert J. Richards, and Michael Ruse, 1st ed., 48–68. Cambridge University Press, 2004. https://doi.org/10.1017/CBO9780511512179.003.

Browne, Janet. 'Charles Darwin as a Celebrity'. *Science in Context* 16:1–2 (2003): 175–194. https://doi.org/10.1017/S0269889703000772.

"Creation", Rotten Tomatoes Website, https://www.rottentomatoes.com/m/1205717-creation.

"Darwin biopic to launch Toronto", *BBC News Online*, July 15, 2009, http://news.bbc.co.uk/1/hi/entertainment/8151456.stm.

"Darwin-Wallace Plaque", *The Times*, July 5, 1958, 8.

Dronamraju, Krishna R., and Joseph Needham. *If I Am to Be Remembered: Correspondence of Julian Huxley*. World Scientific, 1993.

Dunkley, Chris. "Few Goodies at Christmas", *The Financial Times*, January 5, 1983, 13.

Elsdon-Baker, Fern. 'Creating Creationists: The Influence of "Issues Framing" on Our Understanding of Public Perceptions of Clash Narratives between Evolutionary Science and Belief'. *Public Understanding of Science* 24:4 (2015): 422–439. https://doi.org/10.1177/0963662514563015.

Fahy, Declan. *The New Celebrity Scientists: Out of the Lab and Into the Limelight*. Rowman & Littlefield, 2015.

"Five Hundred Million Years", *The Radio Times*, April 18, 1958, 15.

Foster, Patrick. "Eighth Bafta slides into hands of Attenborough for reptile series", *The Times*, April 27, 2009, 9.

Good, Ronald. *Features of Evolution in the Flowering Plants*. Longmans, Green & Co., 1956.

Gouyon, Jean-Baptiste. *BBC Wildlife Documentaries in the Age of Attenborough*. Springer Nature, 2019.

Greene, John C. 'Reflections on the Progress of Darwin Studies'. *Journal of the History of Biology* 8:2 (1975): 243–273.

Gregory, Jane, and Simon Jay Lock, 'The Evolution of "Public Understanding of Science": Public Engagement as a Tool of Science Policy in the UK', *Sociology Compass* 2:4 (2008): 1252–1265, https://doi.org/10.1111/j.1751-9020.2008.00137.x.

Hall, Alexander, Tom Kaden, and Fern Elsdon-Baker, "Darwin Day: Celebrating Without Deifying", *Science and Religion: Exploring the Spectrum Blog*, February 2016. https://sciencereligionspectrum.org/in-the-news/darwin-day-celebrating-without-deifying/.

Hennessy, Elizabeth. 'Mythologizing Darwin's Islands'. In *Darwin, Darwinism and Conservation in the Galapagos Islands: The Legacy of Darwin and Its New Applications*, edited by Diego Quiroga and Ana Sevilla, 65–90. Social and Ecological Interactions in the Galapagos Islands. Cham: Springer International Publishing, 2017. https://doi.org/10.1007/978-3-319-34052-4_5.

———. *On the Backs of Tortoises: Darwin, the Galapagos, and the Fate of an Evolutionary Eden*. Yale University Press, 2019.

Hesketh, Ian. 'The Recurrence of the Evolutionary Epic'. *Journal of the Philosophy of History* 9:2 (2015): 196–219. https://doi.org/10.1163/18722636-12341300.

Hewer, H. R., and Norman Riley, *XVth International Congress of Zoology, London, 16–23 July 1958: Proceedings*, Volume 1, XVth International Congress of Zoology, 1959.

Himmelfarb, Gertrude. *Darwin and the Darwinian Revolution*. Chatto and Windus, 1959.

Hooper, Rowan. "Review: *Creation: The Movie*", *New Scientist*, September 23, 2009, 49.
Huxley, Julian. *Evolution, the Modern Synthesis*. Allen & Unwin, 1942.
Huxley, Leonard. *The Life and Letters of Thomas Henry Huxley*. D. Appleton & Co., 1901.
Johnson, M. L., Michael Abercrombie and G. E. Fogg (Eds), *New Biology 28*, January 1959, Penguin.
Korn, Eric. "remainders", *The Times Literary Supplement*, May 15, 1981, 541.
Lightman, Bernard. 'The Many Lives of Charles Darwin: Early Biographies and the Definitive Evolutionist'. *Notes and Records of the Royal Society* 64:4 (2010): 339–358. https://doi.org/10.1098/rsnr.2009.0059.
Medawar, Peter. *The Future of Man: The BBC Reith Lectures 1959*. Basic Books Inc., 1960.
Moore, James, and Adrian Desmond, *Darwin: The Life of a Tormented Evolutionist*. Penguin, 1992.
"Notes and News." *The Auk*, 75:3 (1958): 378–380.
Penny, David. 'Epigenetics, Darwin, and Lamarck'. *Genome Biology and Evolution*, 7:6 (2015): 1758–1760. https://doi.org/10.1093/gbe/evv107.
Royal Society of London, *Notes and Records*, 14:1, 1959.
Ruse, Michael. 'The Darwin Industry: A Guide'. *Victorian Studies* 39:2 (1996): 217–235.
Scott, Peter. "Our Galapagos Expedition", *The Radio Times*, September 11, 1959, 3.
Scott, Peter, and Philippa Scott, *Faraway Look II*. Cassell, 1960.
Smocovitis, Vassiliki Betty. 'The 1959 Darwin Centennial Celebration in America'. *Osiris* 14 (1999): 274–323.
———. *Unifying Biology: The Evolutionary Synthesis and Evolutionary Biology*. Princeton University Press, 1996.
Sparks, John, and Tony Soper, *Penguins*. David & Charles, 1967.
Tax, Sol, and Charles Callendar, *Evolution after Darwin: the University of Chicago Centennial*, Volume III. University of Chicago Press, 1960.
"The BBC's Darwin Season: marking the life and work of Charles Darwin – introduction", *BBC Press Pack*, January 21, 2009, http://www.bbc.co.uk/pressoffice/pressreleases/stories/2009/01_january/21/darwin.shtml.
"The World of Science: David (Cannon) and Goliath (Darwin)", *The Illustrated London News*, July 5, 1958, 32.
Unsworth, Amy, and David Voas. 'The Dawkins Effect? Celebrity Scientists, (Non) Religious Publics and Changed Attitudes to Evolution'. *Public Understanding of Science*, 25 (2021). https://doi.org/10.1177/0963662521989513.
van Wyhe, John. "1909: The first Darwin centenary", *The Complete Work of Charles Darwin Online*, 2002, http://darwin-online.org.uk/1909.html.

Weindling, Paul. '"Julian Huxley and the Continuity of Eugenics in Twentieth-Century Britain"'. *Journal of Modern European History* 10:4 (2012): 480–499. https://doi.org/10.17104/1611-8944_2012_4.

Wyhe, John van. 'Darwin Online and the Evolution of the Darwin Industry'. *History of Science* 47:4 (2009): 459–473. https://doi.org/10.1177/007327530904700407.

Zakariya, Nasser. *A Final Story: Science, Myth, and Beginnings*. University of Chicago Press, 2017.

Index[1]

[1] Note: Page numbers followed by 'n' refer to notes.

A. Hall, *Evolution on British Television and Radio*, Palgrave Studies in Science and Popular Culture,
https://doi.org/10.1007/978-3-030-83043-4

Printed by Printforce, United Kingdom